Sustainable Production, Life Cycle Engineering and Management

Series editors

Christoph Herrmann, Braunschweig, Germany
Sami Kara, Sydney, Australia

Modern production enables a high standard of living worldwide through products and services. Global responsibility requires a comprehensive integration of sustainable development fostered by new paradigms, innovative technologies, methods and tools as well as business models. Minimizing material and energy usage, adapting material and energy flows to better fit natural process capacities, and changing consumption behaviour are important aspects of future production. A life cycle perspective and an integrated economic, ecological and social evaluation are essential requirements in management and engineering. This series will focus on the issues and latest developments towards sustainability in production based on life cycle thinking.

More information about this series at http://www.springer.com/series/10615

Sebastian Thiede · Christoph Herrmann
Editors

Eco-Factories of the Future

Springer

Editors
Sebastian Thiede
Institute of Machine Tools and Production Technology
Technical University Braunschweig
Braunschweig, Germany

Christoph Herrmann
Institute of Machine Tools and Production Technology
Technical University Braunschweig
Braunschweig, Germany

ISSN 2194-0541 ISSN 2194-055X (electronic)
Sustainable Production, Life Cycle Engineering and Management
ISBN 978-3-030-06711-3 ISBN 978-3-319-93730-4 (eBook)
https://doi.org/10.1007/978-3-319-93730-4

This Springer imprint is published by the registered company Springer Nature Switzerland AG
The registered company address is: Gewerbestrasse 11, 6330 Cham, Switzerland

Series Editors' Foreword

Sustainability has become an important issue for manufacturing companies within the last years. This includes three closely connected dimensions—economy, environment and society. Legislative issues, energy/resource prices and availability as well as customer and general expectations clearly underline the strong interdependencies of economic and environmental goals when planning and operating factories. Companies strive towards both lean and "green" factories, ideally with an integrated approach seamlessly bringing together both target dimensions. They also increasingly realise the potential of addressing environmental issues as drivers for product and process innovations and the opportunity for differentiation in the global competition.

In the last decade, numerous contributions towards sustainable factories were published in academia and industry—also in our series "Sustainable Production, Life Cycle Engineering and Management". Those approaches typically address specific analyses and improvements in designated fields of action, e.g. energy efficiency of production machines and processes. Obviously, a factory consists of various constituting elements and, thus, there exist many different possibilities towards improving factories.

As a result, this book gives a compact overview of selected solutions on different factory levels based on a common factory framework. This will create awareness of different options towards more sustainable factories while also emphasising the need of a systemic perspective. Contributions from different international authors, from academia and industry, were compiled. The solutions contributed by leading research institutions and companies have been mostly developed in large EU or nationally funded research projects and were tested in industrial pilot projects across Europe. Besides technological solutions, also methodological approaches are presented, which cover topics such as factory planning, manufacturing simulation, energy management or life cycle evaluation of production systems. This will be another invaluable addition towards promoting sustainability in manufacturing.

Sydney, Australia
Sami Kara
The University of New South Wales

Braunschweig, Germany
Christoph Herrmann
Technische Universität Braunschweig

Preface

Without a question, the generation of valuable goods is the main purpose of industrial companies, and manufacturing is still a cornerstone of a positive development and prosperity of countries. Traditionally, the focus of improvement activities is on manufacturing goods in defined quantities and qualities for the lowest possible costs and in a short time.

In the last decades, the environmental perspective strongly grew in importance and added new goals that need to be dealt with. Originally strongly driven by "end of pipe" approaches and typically by separate departments, the main focus was on avoiding emissions or treatment of waste in order to cope with legislative requirements. Nowadays, producing companies have realised the extended and economically relevant scope of addressing environmental issues. Thus, a more proactive perspective moves into the spotlight, trying to integrate topics such as energy and resource efficiency into the daily production business. New dynamics on the energy market such as the increasing integration of volatile renewable energy sources in the energy grid lead to even more challenges, but also potentials in this context. Nowadays, not only the efficient and also the flexible utilisation of energy carriers are of both economic and environmental relevance. Recent trends such as digitalisation open up new ways to deal with those aspects through innovative approaches for metering, data analytics, simulation as well as decision support systems and control approaches. However, the question arises where benefits can be achieved through those approaches, while additional efforts and expertise might be necessary.

To realise those benefits, an integrated perspective is necessary which enables companies to balance different target criteria and strive towards "Eco-Factories of the Future". Therefore, it is important to pursue a holistic factory perspective which takes into account all factory elements—production machines, technical building services and the building shell—and their dynamic interdependencies.

Against this background, this book presents technologies as well as digital methods and tools which help to foster future eco-factories. After setting an overall framework, the different chapters cover various topics from the perspective on single machines and process chains, technical building services (e.g. compressed air, energy storage) and the factory as a whole. The editors sincerely want to thank all involved authors from industry and academia for their valuable contributions, which completed this comprehensive and interdisciplinary perspective on eco-factories of the future.

Braunschweig, Germany

Christoph Herrmann
Sebastian Thiede

Contents

Chapter 1
Towards Eco-Factories of the Future

Denis Kurle, Sebastian Thiede and Christoph Herrmann

Abstract One-third of the global greenhouse gas emissions is caused by combusting fossil fuels to manufacture products and goods. Contemporary initiatives to lower the emissions chiefly focus on eco-efficiency, seeking for minimized energy demand and to a smaller extend also for minimized resource consumption. In addition to that, new technologies, a demographically changing workforce and the desire for new individualized products put further pressure on manufacturing companies. To encounter all those new trends and challenges successfully, various perspectives of a factory have been proposed to improve the understanding of all involved interdependencies between the factory elements. However, most of those perspectives lack to take all contemporary challenges and trends into account. Therefore, this chapter provides an extended perspective on the elements of a future factory.

1.1 Introduction

Various examples from history substantiate the important role of manufacturing for the development and prosperity of nations and hence also the growing wealth of the population (Gutowski et al. 2013; Reinert 2007). A key part for that development plays the machinery industry. Since the manufacturing of goods needs adequate production machinery including different components made for specific machine tools to produce in turn parts for diverse goods, manufacturing has strongly contributed to the vast economic growth of the last two hundred years (Rynn 2011). The importance of manufacturing can be also derived by comparing, for example, the world trades allocated to merchandise export (US$ 18.3 trillion) with the amount of commercial services (US$ 4.3 trillion) (World Trade Organization 2013). Based on that, a country's trade balance often reveals information about its wealth in two ways since

D. Kurle · S. Thiede (✉) · C. Herrmann
Sustainable Manufacturing & Life Cycle Engineering Research Group,
Institute of Machine Tools and Production Technology (IWF),
Technical University Braunschweig, Braunschweig, Germany
e-mail: s.thiede@tu-braunschweig.de

S. Thiede and C. Herrmann (eds.), *Eco-Factories of the Future*, Sustainable Production, Life Cycle Engineering and Management,
https://doi.org/10.1007/978-3-319-93730-4_1

services often depend on manufactured goods and cannot be offered independently (Aurich and Clement 2010; Kuntzky 2013).

In the light of the outlined importance of manufacturing related to a country's development and wealth, it is a country's objective to sustain or re-establish a strong manufacturing industry. However, to achieve this factories have to adapt to new and evolving challenges in manufacturing to remain competitive. This requires a responsible dealing with those challenges, which primarily involve among others:

- **Digitalization** as an industrial and societal challenge affects many areas of work and private life. From a manufacturing perspective, particularly more mature and economically feasible technologies in conjunction with evolving technologies and approaches from the Industry 4.0 umbrella offer significant potentials for improvements (Kagermann et al. 2013).
- The advancing **urbanization** process poses a challenge for industrial societies, since 66% of all people are expected to live in cities in 2050 (acatech 2016). This means that urbanization is of particular importance in industrialized countries, while cities are also drivers for the depletion of new knowledge and therefore offer the best prerequisites for disruptive innovations in industry as well as society.
- This comes along with a people's increasing desire for **individualization**. As a result, higher product varieties with lower specific lot sizes follow, which implies a rethinking of current manufacturing practices for instance based on economies of scale.
- As a consequence of the aforementioned challenges, a **change of work** towards an increasingly digitized working World 4.0 requiring new skills, qualifications and knowledge will follow. Although many effects of the digitized working World 4.0 are not regarded as determinable, they are often considered to be designable due to its often interdisciplinary context (Bundesministerium für Arbeit und Soziales 2016; Bundesministerium für Bildung und Forschung 2016; Hirsch-Kreinse 2014; United Nations 2014). This challenge of qualifying employees is further reinforced by the demographic change in societies.
- Another challenge relates to **sustainability** aspects in manufacturing, which is associated to a lack of natural resources and therefore prices and legislative regulations to comply with in terms of greenhouse gas or volatile organic compounds emissions as well as many others. As a result, energy and resource efficiency improvements have been and still are high on the agenda of manufacturing companies. In that context, the matching of factory internal energy and media demands with external volatile renewable energy sources grows in importance for manufacturing companies and is summarized as energy flexibility (Duflou et al. 2012).

The complex influence of those challenges on manufacturing companies implies that many of them cannot be viewed in isolation from each other to meet customer needs on the one hand and comply with emerging market situations on the other hand (e.g. in product volume per variant). As a result, the contemporary situation for the manufacturing industry is on a verge of change, which can be best understood by looking at its evolution over the last two centuries. This evolution involves several paradigm changes from Craft Production over Mass Production to Lean Manufactur-

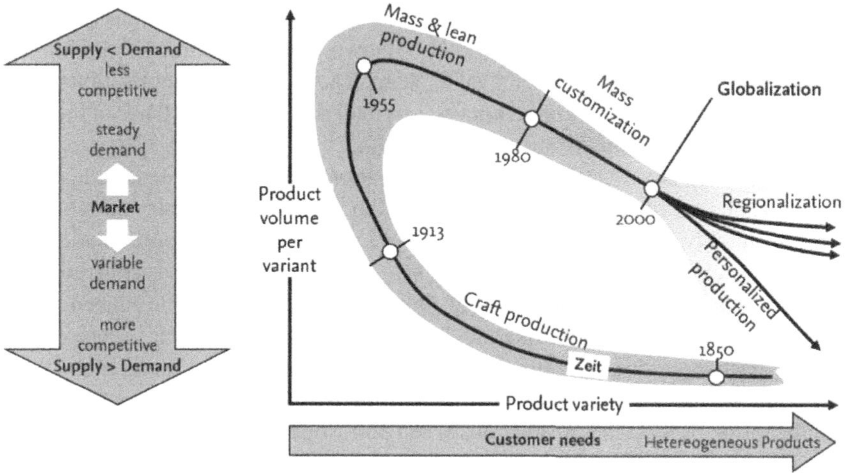

Fig. 1.1 Relationship between volume and variety in manufacturing paradigms (reproduced from reference Koren 2010 with permission)

ing, Mass Customization, while it is still ongoing as manufacturing companies are in the process of adopting to new challenges and trends (Hu et al. 2011). Figure 1.1 illustrates the evolution of the aforementioned manufacturing paradigms, which are described in more detail in the following.

The first paradigm is known as *craft production* since it responded to a specific customer order. The manufactured products were unique, which implies a high product variety through a high flexibility. However, since all products were created manually, they incurred relatively high costs as no automated manufacturing systems were in place for this paradigm (Hu 2013).

Mass production denotes the second paradigm, which fosters the production of products at lower cost through large-scale manufacturing. However, there was only a limited variety of products applied. Typical characteristics of this paradigm were Henry Ford's moving assembly line and his statement: "Any customer can have a car painted any color that he wants so long as it is black" (Ford and Crowther 1922). The peak of the mass production paradigm was reached after World War II, since demands for products increased rapidly. In general, mass production comprises three different principles. The first one regards interchangeability to improve the assembly selection of parts. The second principle denotes the concept of a moving assembly line, which transports the products, e.g. cars, to the workers. Since the workers executed the same tasks repeatedly, the products could be manufactured with lower tolerances, quicker and therefore at lower assembly costs (EyeWitness to History 2005). The third principle describes what is commonly known as Taylorism. With the help of Taylorism, work is subdivided into several specialized but repetitive tasks.

Lean manufacturing also emerged as a result from World War II in Japan due to the country's limited resources and is often referred to as the "Toyota Production

System" as its major developer. From a manufacturing perspective, this paradigm aims at minimizing all sorts of waste—also known as *Muda*—through maximizing customer value along the product's value chain. The widespread application of this paradigm in contemporary production systems underlines the importance of the Lean Management Philosophy (Womack et al. 1990).

Mass customization originated from the customer demand for increased product variety at the end of the 1980s. Particularly, the automotive industry adopted that principle quickly, leading, for example, to 1017 different possible combinations for the BMW 7 series alone (BMW Group 2014). Again, three principles constitute this paradigm (Hu 2013). At first, different family architectures are introduced which differentiate between product modules that are shared among each product while others are only provided individually upon customization request (Tseng et al. 1996). This helps to manifest the second principle, which describes the distinction of the manufacturing process into a rather push- and pull-oriented production and assembly part. This implies a delayed differentiation of products, which is typically known as a customer decoupling point to customize the final product while saving costs and increasing the responsiveness of the assembly system (Lee and Tang 1997; Ko and Hu 2008). The reconfigurability of a manufacturing system states the third principle and denotes the prerequisite for changing product mixes and demands to improve the overall system's performance (Koren et al. 1999; Koren et al. 1998). This, however, comes at the expense of handling a manufacturing system with an increased complexity (Hu et al. 2008).

As a result from smaller product volumes and a rising variety (see Fig. 1.1), it emerges a trend towards more *personalized products* (Hu 2013) Thus, various different work steps and cycle times are involved which are often independent from typical manufacturing KPIs, such as tact times of typical assembly lines. Ultimately, this may lead to the economically feasible manufacturing of very small lot sizes as a gain from improved production system flexibility.

Another arising trend is the need for environmentally benign manufacturing due to higher living standards, a growing demand of consumers for products and consequently more resources for energy production, which lead to rising energy and resource prices (Herrmann 2010). Stricter legislative regulations to comply with emission boundaries follow, which is why a *sustainable production* is more important than ever for manufacturing companies.

Furthermore, *information and communication technology (ICT)* will play an important role in the future manufacturing sector. This is already reflected by the past progression: Starting with first NC machines in the 1950s (Pease 1952) over computer-controlled production cells in the late 1960s (Talavage and Hannam 1987) to the idea of computer-integrated manufacturing (CIM) in the late 1970s (Mitchell 1991) a constant advancement is notable. Since the costs of ICT components have dropped significantly while the technology becomes more versatile and powerful, there will be multiple new ways how these components will change future production systems, as indicated by the present research trends towards cyber-physical systems (CPS) (European Commission 2014).

The aforementioned trends bear also a *social influence* on future manufacturing companies in multiple ways. At first, professionals will have to learn how to cope with the new trends through *production-related learning environments* to familiarize themselves with the new technologies and circumstances (DIHK 2011; VDI Nachrichten 2014; Damodaran 2001). Secondly, personal desires gain in importance so that living circumstances, for example in the light of a progressing urbanization, will further reinforce the wish for short commuting distances (The World Bank 2013). Hence, it remains questionable whether factories and therefore production systems might be integrated into residential areas as good neighbours (Chertow 2007). Thinking and taking this idea one step further, *future production ecosystems*, may be considered as a place which is not only ecological efficient in itself but also enables benefits for its environment from a holistic perspective. In that context, factories may be in a symbiosis with their surroundings in terms of their role in smart grids, local (climate) constraints or sharing conditioning and treatment units e.g. for fresh, process or wastewater with and from residential areas (Chertow 2007; Greenpeace 2012).

1.2 Factories of the Future Framework

To respond to the outlined trends such as urbanization or digitalization in form of cyber-physical production system (CPPS), Herrmann and colleagues propose a renewed holistic understanding of a factory of the future (Herrmann et al. 2014). This understanding seeks to address more than ever all three dimensions of sustainability—economy, ecology and society. While aiming at higher profitability from an economic perspective, the factory should also be associated as a place with a positive impact on its surrounding area by improving the quality of air and water, exploiting local waste flows, providing renewable energies or by acting as storage for surplus. From a social perspective, future factories should be perceived as a people's place, where they can focus on collaborative learning and developing new skills. Figure 1.2 shows this new perspective and differentiates between four main aspects worth considering in the planning of future factories which will be further explained in the following:

- Symbiotic flows and urban integration of the factory
- Adaptable factory elements: adaptive building shell, modular and scalable TBS, and flexible production system
- Production cloud and cyber-physical systems
- Learning and training environments.

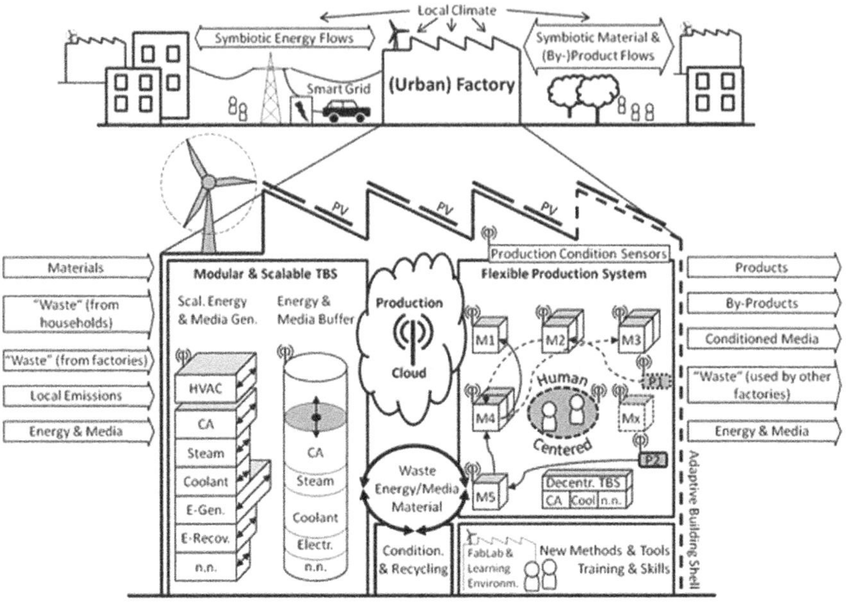

Fig. 1.2 Holistic understanding of a factory of the future and its flows (Herrmann et al. 2014)

1.2.1 Symbiotic Flows and Urban Integration of the Factory

While many contemporary approaches of sustainable production are chiefly concerned about minimizing resource and energy input flows as an efficiency effort (Despeisse et al. 2012; Braungart et al. 2007), this may not be a remedy for an adequate long-term solution. This is because efficiency efforts describe strategies for damage management as opposed to strategies that rather focus on new shifts in strategy. Such a shift could imply for future factories to cease opportunities for a potential upcycling of material or media quality over time. Consequently, factories will be capable of generating a positive recoupling between the two sustainability dimensions of economy and ecology, which is known as eco-effectiveness. Thus, eco-effectiveness helps to close loops between material and energy flows within the factory itself but also in collaboration with the factory's external environment (Braungart et al. 2007). Positive effects may entail that solid waste will be used for new products while wastewater will be treated and renewable energies will be produced or stored helping to neutralize local emissions.

An eco-effective production resembles strong similarities to a biological system, where material flows serve as biological nutrients for living systems while "waste" does not exist, as it is an input for another living system (Ayres and Simonis 1994). This idea is also adopted by the concept of a factory of the future as it incorporates a symbiotically sharing of material and energy flows with its (urban) infrastructure and/or with other factories to foster the integration of smart grid technologies. Some examples of this idea have already been rudimentarily implemented in form of so-called industrial symbioses within eco-industrial parks, such as Kalundborg or the Haven in Rotterdam. Chertow gives an overview of further realized projects in the context of industrial symbioses (Chertow 2000).

1.2.2 *Adaptable Factory Elements*

The literature presents a variety of perspectives and slightly different terms regarding the structure, composition and elements of a factory (Wiendahl et al. 2007; Müller et al. 2009). However, among those perspectives there is a consensus to group the elements of a factory into the building shell, the technical building services (TBS) and the production equipment, which include single machines or process chains (Hesselbach 2012; Thiede 2012). This categorization helps to emphasize the manifold connections (e.g. regarding production system utilization, machine states, media demands, weather) between all involved factory elements and its entities with each other. This connection is realized through different material, energy, media and information flows, which have to be maintained and controlled subject to induced internal or external changes on the production system (Westkämper and Zahn 2009). Thus, a factory and its production system can also be perceived as a complex control system with internal and external influencing variables (Thiede 2012). To transform factories, it is important to bear the following five principles in mind while designing future factory elements: *modularity, scalability, universality, compatibility and mobility* (Wiendahl et al. 2007).

1.2.2.1 Adaptive Building Shell

The building shell mainly insulates the factory and separates the other factory elements from the outer conditions and the local climate, respectively. Future requirements concerning the building shell regard their potential of flexibility in terms of changeability. As the whole factory is chiefly determined by its primary structure, e.g. the distance between the supporting columns of the building, it seems beneficial to use larger distances of 30–40 m between two pillars instead of 20 m, which is state of the art at the moment. This helps to gain spatial flexibility and therefore higher degrees of freedom regarding utilization for only little increased invests by approximately five per cent (Wiendahl et al. 2005). Since the structure of a factory is commonly planned upfront, the building shell and its pillars are hardly changeable afterwards. Thus, flexibility limits can be extended by adopting those new principles in the early planning stage or by thinking about completely new building concepts, which are more flexible to changing requirements. Examples of such new building concepts and structures are factories comprising modular containers, factories within transportation systems such as ships or trains as well as air-inflated factory structures (Wiendahl et al. 2005).

From an ecological perspective, the building shell can contribute to either being made of appropriate building materials, such as cement, or by incorporating useful functions, e.g. for cleaning the air by absorbing and photocatalytically decomposing various pollutants from the air (Chen and Poon 2009). The roof of a factory is predestined for the installation of renewable power plants to acquire and save energy from solar and wind. To support the power generation new materials with integrated energy generation capabilities like building-integrated photovoltaics (BIPV) and superseding retrofit solar power systems can be employed (Eiffert and Kiss 2000). Provided that enough energy can be stored within the factory, this could enable energy-self-sufficient factories, which can act independently from the grid and may be a favourable solution for rural areas of developing countries.

1.2.2.2 Modular and Scalable TBS

In production systems exist diverse media flows which are directly linked to the manufacturing operations. Many of those media flows are provided by TBS, which pre-process energy and media flows including compressed air, heating, cooling, water and coolants subject to the required production conditions (Thiede 2012). In future factories, TBS will be linked both physically and virtually to the production systems. This will enable to go beyond mere production monitoring purposes through an automated optimization production condition concerning controlling variables such as temperature, lighting, humidity.

Apart from factory internal energy and media provision, the TBS technologies may also be used to integrate factory external energy and media demands and supplies to interact symbiotically with the factory surroundings. As a positive effect from this potential collaboration, future factories may be used as energy and media buffers to level fluctuating inputs and outputs in certain areas and thus help stabilizing, for instance, the electric grid as a backup for power blackouts (Eyer and Corey 2010; Masaud et al. 2010). With respect to the realization of factory internal storage options, central battery systems or compressed air energy storages as well as decentralized systems such as electric vehicles or even products, which imply an increased production in times of energy availability, appear to be reasonable.

To be able to incorporate such changes resulting from the integration of surrounding energy and media flows as well as renewable energy technologies, future TBS must be designed and controlled to react quickly on fluctuating demands and supplies. In addition to that, fluctuating customer orders and a high degree of product individualization leading to more fragmented process chains with unequal energy and media demands further emphasize this need of quick TBS responsiveness. Because of decreasing product life cycles, more frequent changes of the production system can be expected, which will also entail an impact on the TBS. Therefore, a modular, expandable and partly decentralized TBS structure could be more suitable than traditional centralized systems, although it may come at the expense of lower a degree of efficiency of small or decentralized systems compared to larger, monolithic TBS structures. The design of decentralized TBS modules may be similar to the idea of plug-and-produce, which has been derived from the plug-and-play principle of computing (Hildebrand et al. 2005). In that sense, the high degree of hardware and software standardization facilitates the integration of additional elements and devices by simply plugging them to machines, transport systems or TBS and start working instantly, similar to an USB device connected to a PC. Consequently, increased demands of, for example, compressed air could be handled with a plug-and-produce extension for the existing compressor, which therefore owns standardized interfaces. This could help to avoid an over-dimensioning of systems as upgrading and down-grading would be possible at all times with minor efforts.

1.2.2.3 Flexible Production System

A flexible production system of a future factory is capable of responding to an increased variety and complexity of future products. To operate such a system profitably while using minimal resources, a high degree of machine utilization is required. However, particularly for assembly operations, traditional concepts are not able to maintain an optimal utilization of machines while producing components with a high variance in processing times on the same production line (Krajewski et al. 2010) Since all product variants require a different work content, varying processing times and hence cycle times for all products and processes follow. Thus, traditional production and assembly lines seem to be not suitable to respond to future requirements in terms of flexible production and assembly. A promising approach to alleviate this issue could be an increased versatility of the production machines. This allows every machine to assemble a big variety of different product variants. As a consequence, products may choose different paths through the production system subject to their preferred objective, which could be a time, priority or energy optimized path for each individual product variant. The preferred paths for each product further depend on the properties of the other product variants currently in the system as well as the individual states of the machines in the multi-machine production environment. To implement such a flexible production system, an appropriate flexible transport system, allowing an individual distribution of material to the production resources, is needed as well.

Similar to the TBS, a flexible production system may also follow the plug-and-produce philosophy (Hildebrand et al. 2005) to provide a high degree of scalability to respond to changing market demands. Such a system can be well represented by multi-agent system (MAS) approaches, which base upon a set of distributed autonomous but cooperative entities such as products and machines (Wooldridge 2002). In that sense, many decisions then can be made decentralized and collaboratively by the agents, based on their individual logic, resulting in an optimized performance of the whole production system. This approach also facilitates a quick reconfigurability of the system, because no fixed central systems for production planning and control have to be changed. As appropriate information flows are essential for operation, most elements of the system will be equipped with sensors and empowered to communicate with each other about their current states and potential orders or product queues waiting in suitable buffers.

Apart from all technical aspects, it is important for future production systems to be designed around human employees as they will be needed in all phases of factory operations, from the planning through operation to maintenance and repair (Zuehlke 2010). Furthermore, technologies should be designed and developed in such a way so that human cognitive and sensomotor abilities can be used and/or systematically developed, e.g. using learning and training environments (see Sect. 1.2.4). Thus, technology has to be adopted to human needs instead of a conversely adaption of human action to technological constraints.

1.2.3 Production Cloud and Cyber-Physical Systems

Since information technology (IT) will further increase overall factory layers, including a wide application of sensors on shop floor level, the collection, consolidation and processing of those information will be very important. As monolithic IT architectures will fade, there will be space for new, agile platforms, which can gather data and use them to provide different services and/or make decentralized decisions. Independent entities or agents can then perform those services or decisions. For example, CPS can support human workers on shop floor level by providing context-specific live information gathered by the sensors from the production system, which will be pre-processed for the specific task.

The vast amount of gathered data in the factory of the future requires the setup of a decentralized data pool, the "production cloud", which collects and allocates all information from the production system as well as TBS entities. Typical data include productivity (e.g. processing and cycle times), energy and resources data (e.g. energy demands per process/machine/part), product-related information (e.g. product quality indicators) or data concerning the production conditions (temperature, humidity, lighting etc.) as well as the current status and state of all factory elements and its entities. Having all those data stored and accessible in a production cloud entails several benefits, such as a high degree of transparency through embedded monitoring and control functions, which provides useful information for future improvement measures. The production cloud could also serve as a link to cloud manufacturing services (Tao et al. 2011), which appears to be an adequate approach for the manufacturing of individualized products within virtual manufacturing networks. Every factory of the network can offer its services with regard to free capacity, deliverable quality or achievable delivery times, causing individual supply chains for every product or variant. This idea is very similar to the concept of an Internet of things (IoT), which was presented in the 1990s (Weiser 1991; Ashton 2009). According to the IoT, all physical elements also have a virtual representation in an Internet resembling structure. To achieve such a virtual representation, a broad installation of sensors, actuators and processors including distinct identifiers of all objects, e.g. realized by integration of RFID tags, are required to duplicate the physical system and its behaviour. Such a vision of a Factory of Things, composed of self-organizing smart objects without a traditional hierarchy, is also known as Smart Factory (Zuehlke 2010).

1.2.4 Learning and Training Environments

Regardless of all technological progress, human abilities will still be a significant success factor of future factories. Therefore, highly educated employees are required which undergo continuous training to keep pace with changing external requirements and technological improvements. To facilitate this learning and training process, "Learning Factories" (also referred to as Teaching Factories) could play an important role in educating the employees since they can experiment and research new topics at realistic production processes (Chryssolouris et al. 2013). Those processes can typically be adjusted for case-specific teaching purposes of different target groups. Furthermore, such learning environments also allow for the communication and testing of theoretical knowledge in practical application and learning results can be transferred to real factory applications. A Learning Factory can comprise both physical and digital environments. Physical environments contain real system components such as machines, assembly, logistics, information and energy flow modules, while the digital environments include planning, modelling, visualization and simulation tools to experience different effects and measures virtually. Physical and digital environments can also work together, where some improvements scenarios can be first modelled and evaluated virtually and subsequently be tested and assessed in the physical learning environment (Wagner et al. 2012). The concept of Learning Factories is an established concept in academic research, but also an increasing implementation of Learning Factories and scenarios can be observed in industry.

Future factories may involve fabrication laboratories or "FabLabs", which contain a collection of tools for design and modelling, prototyping and fabrication, instrumentation, testing, debugging and documentation for a wide range of production relevant applications (Gershenfeld 2012; Mikhak et al. 2002). This will empower people to design, physically realize and test their own personalized product creations in the FabLab by using the provided infrastructure and machines. Since this enables an easy entrance to new technologies, this approach may be especially beneficial for regions with lower education or prosperity, as they support regional development by catering specific individual needs of the people. In addition to that, FabLabs also state a platform for knowledge transfer to broad parts of the society in industrial countries (The International Fab Lab Association 2012).

1.3 Conclusion and Outlook

Manufacturing constitutes an important factor for the prosperity of nations and a vital source of innovation and development. To guarantee the competitiveness of the manufacturing industry, factories had and always have been under pressure to adapt to new challenges and trends. This resulted in several changes of manufacturing paradigms over the last two centuries. Today and future factories face various evolving trends. Those trends involve highly customized products manufactured by an eco-friendly production system with significantly lower environmental impacts. This is often realized using new ICT technologies while providing social options for production-related learning and an integration of the factories into its spatial (urban) surrounding.

To face those trends, this chapter presents a holistic factory perspective of the future, which addresses the contemporary trends in manufacturing. The presented vision contains four main aspects: first, a symbiotic integration of the factory into its surroundings and in particular to urban or domestic areas as well as an orientation towards eco-effectiveness, heading for a positive recoupling between economy and ecology. Second, since a high degree of flexibility of the factory infrastructure is needed, adaptable and modular main factory elements, namely building shell, TBS and a flexible production system, are needed. Third, as the operation of such a system is only feasible through a decentralized gathering and use of information, the integration of a production cloud concept is proposed. Furthermore, a widespread application of ICT is indispensable, which will help to link the physical and digital factory representations with each other. The fourth and last aspect addresses the necessity to provide learning and training to hone human skills and abilities as future production system will be still human-centred. To achieve that, the implementation of specific learning environments such as Learning Factories seems promising.

References

Ashton K (2009) That 'internet of things' thing. RFiD J 22:97–114

Aurich JC, Clement MH (2010) Produkt-Service Systeme. Gestaltung und Realisierung. Springer, Berlin, Heidelberg

Ayres RU, Simonis UE (1994) Industrial metabolism: restructuring for sustainable development. United Nations University Press, Tokyo

BMW Group (2014) Wie ein Fahrzeug entsteht. http://www.bmwgroup.com/bmwgroup_prod/d/0_0_www_bmwgroup_com/produktion/fahrzeugfertigung/automobilfertigung/erlebnis_produktion/prod_prozesse.shtml. Accessed 20 Apr 2014

Braungart M, McDonough W, Bollinger A (2007) Cradle-to-cradle design: creating healthy emissions—a strategy for eco-effective product and system design. J Clean Prod 15(13–14):1337–1348

Chen J, Poon C (2009) Photocatalytic construction and building materials: from fundamentals to applications. Build Environ 44:1899–1906

Chertow MR (2000) Industrial symbiosis: literature and taxonomy. Annu Rev Energy Env 25:313–337

Chertow MR (2007) Uncovering industrial symbiosis. J Ind Ecol 11(1):11–30

Chryssolouris G, Mavrikios D, Mourtzis D (2013) Manufacturing systems: skills & competencies for the future. Procedia CIRP 7:17–24

Damodaran L (2001) Human factors in the digital world enhancing life style—the challenge for emerging technologies. Int J Hum Comput Stud 55:377–403

Despeisse M, Ball PD, Evans S, Levers A (2012) Industrial ecology at factory level—a conceptual model. J Clean Prod 31:30–39

Deutsche Industrie- und Handelskammertag e.V., DIHK Mittelstandsreport. Sommer 2011 (2011). http://www.dihk.de/

Duflou JR, Sutherland JW, Dornfeld D, Herrmann C, Jeswiet J, Kara S, Hauschild M, Kellens K (2012) Towards energy and resource efficient manufacturing: a processes and systems approach. CIRP Ann Manuf Technol 61(2):587–609

Eiffert P, Kiss GJ (2000) Building-integrated photovoltaic designs for commercial and institutional structures: a sourcebook for architects. DIANE Publishing

European Commission (2014) Call for factories of the future: process optimisation of manufacturing assets. http://ec.europa.eu/research/participants/portal/desktop/en/opportunities/h2020/topics/2180-fof-01-2014.html. Accessed 20 Apr 2014

Eyer J, Corey G (2010) Energy storage for the electricity grid: benefits and market potential assessment guide, Sandia National Laboratory, Report No. SAND2010-0815

Ford H, Crowther S (1922) My life and work. Doubleday, Page & company

Gershenfeld N (2012) How to make almost anything: the digital fabrication revolution. Foreign Aff 91(6):43–57

Greenpeace (2012) Energy [r]evolution: a sustainable world energy outlook, report 4th edition 2012 world energy scenario. http://www.greenpeace.org/international/Global/international/publications/climate/2012/Energy%20Revolution%202012/ER2012.pdf

Gutowski TG, Allwood JM, Herrmann C, Sahni S (2013) A global assessment of manufacturing: economic development, energy use, carbon emissions, and the potential for energy efficiency and materials recycling. Annu Rev Environ Resour 38:81–106 (Palo Alto, CA)

Henry Ford Changes the World, 1908, EyeWitness to History (2005). www.eyewitnesstohistory.com

Herrmann C (2010) Ganzheitliches life cycle management. Nachhaltigkeit und Lebenszyklusorientierung in Unternehmen. Springer, Berlin, Heidelberg, p. 30

Herrmann C, Schmidt C, Kurle D, Blume S, Thiede S (2014) Sustainability in manufacturing and factories of the future. Int J Precis Eng Manuf Green Technol 1(4):283–292

Hesselbach J (2012) Energie- und klimaeffiziente Produktion. Springer-Vieweg, Heidelberg. Figures used with kind permission of Springer Science+Business Media

Hildebrand T, Mäding K, Günther U (2005) Plug + produce: Gestaltungsstrategien für die wandlungsfähige Fabrik, IBF, Chemnitz

Hirsch-Kreinse H (2014) Wandel von Produktionsarbeit – Industrie 4.0. WSI-Mitteilungen (6):S. 421–429

Hu JS (2013) Evolving paradigms of manufacturing: from mass production to mass customization and personalization. Procedia CIRP 3–8

Hu SJ, Zhu X, Wang H, Koren Y (2008) Product variety and manufacturing complexity in assembly systems and supply chains. CIRP Ann Manuf Technol 57(1):45–48

Hu SJ, Ko J, Weyland L, ElMaraghy HA, Lien TK, Koren Y, Bley H, Chryssolouris G, Nasr N, Shpitalni M (2011) Assembly system design and operations for product variety. CIRP Ann Manuf Technol 60(2):715–733

Kagermann H, Wahlster W, Helbig J (2013) Recommendations for implementing the strategic initiative INDUSTRIE 4.0, Final report of the Industrie 4.0 Working Group, acatech. http://www.plattform-i40.de/sites/default/files/Report_Industrie, 204

Kompetenzentwicklungsstudie Industrie 4.0. Erste Ergebnisse und Schlussfolgerungen, Hg. v. acatech - Deutsche Akademie der Technikwissenschaften, München (2016). http://www.acatech.de/fileadmin/user_upload/Baumstruktur_nach_Website/Acatech/root/de/Publikationen/Kooperationspublikationen/acatech_DOSSIER_Kompetenzentwicklung_Web.pdf. Accessed 11 Jan 2017
Ko J, Hu SJ (2008) Balancing of manufacturing systems with complex configurations for delayed product differentiation. Int J Prod Res 46(15):4285–4308
Koren Y (2010) The global manufacturing revolution: product-process-business integration and reconfigurable systems (Vol. 80). John Wiley & Sons.
Koren Y, Hu SJ, Weber TW (1998) Impact of manufacturing system configuration on performance. CIRP Ann Manuf Technol 47(1):369–372
Koren Y, Jovane F, Heisel U, Moriwaki T, Pritschow G, Ulsoy AG, Van Brussel H (1999) Reconfigurable manufacturing systems. CIRP Ann Manuf Technol 48(2):6–12
Krajewski LJ, Ritzman LR, Malhotra MK (2010) Operations management, processes and supply chains. Pearson, Upper Saddle River
Kuntzky K (2013) Systematische Entwicklung von Produkt-Service Systemen, Technische Universität Braunschweig, Vulkan-Verlag, Essen
Lee H, Tang C (1997) Modelling the costs and benefits of delayed product differentiation. Manag Sci 43(1):40–53
Masaud TM, Lee K, Sen PK (2010) An overview of energy storage technologies in electric power systems: what is the future? In: North American Power Symposium, Arlington, TX, USA, pp 1–6
Mikhak B, Lyon C, Gorton T, Gershenfeld N, McEnnis C, Taylor J (2002) Fab Lab: an alternate model of ICT for development. In: 2nd international conference on open collaborative design for sustainable innovation
Mitchell FH (1991) CIM systems: an introduction to computer-integrated manufacturing. Prentice Hall
Müller E, Engelmann J, Löffler T, Strauch J (2009) Energieeffiziente Fabriken planen und betreiben. Springer, Berlin, Heidelberg
Pease W (1952) An automatic machine tool. Sci Am 187(3):101–115
Reinert ES (2007) How rich countries got rich … and why poor countries stay poor. Constable, London
Rynn J (2011) Six reasons manufacturing is central to the economy. http://www.rooseveltinstitute.org/new-roosevelt/six-reasons-manufacturing-central-economy. Accessed 20 Apr 2014
Talavage J, Hannam RG (1987) Flexible manufacturing systems in practice: design: analysis and simulation. CRC Press, pp 45–48
Tao F, Zhang L, Venkatesh VC, Luo YL, Cheng Y (2011) Cloud manufacturing: a computing and service-oriented manufacturing model. Proc Inst Mech Eng Part B J Eng Manuf
The International Fab Lab Association (2012) The Fab Charter. http://www.fablabinternational.org/fab-lab/the-fab-charter
The World Bank (2013) Developing countries need to harness urbanization to achieve the MDGs: IMF-World Bank report. http://www.worldbank.org/en/news/press-release/2013/04/17/developing-countries-need-to-harness-urbanization-to-achieve-mdgs-imf-world-bank-report. Accessed 20 Apr 2014
Thiede S (2012) Energy efficiency in manufacturing systems. Springer, Berlin u.a.
Tseng MM, Jiao J, Merchant ME (1996) Design for Mass customization. CIRP Ann Manuf Technol 45(1):153–156
VDI Nachrichten (2014) Aus Sicht der Industrie ist der Fachkräftemangel spür- und sichtbar. http://www.vdi-nachrichten.com/Management-Karriere/Aus-Sicht-Industrie-Fachkraeftemangel-spuer-sichtbar. Accessed 20 Apr 2014
Wagner U, AlGeddawy T, ElMaraghy H, Müller E (2012) The state-of-the-art and prospects of learning factories. Procedia CIRP 3:109–114
Weiser M (1991) The computer for the 21st century. Sci Am 265(3):94–104

Weißbuch Arbeit 4.0. Bundesministerium für Arbeit und Soziales (2016) Berlin. http://www.bmas.de/SharedDocs/Downloads/DE/PDF-Publikationen/a883-weissbuch.pdf. Accessed 11 Jan 2017
Westkämper E, Zahn E (2009) Wandlungsfhige Produktionsunternehmen: Das Stuttgarter Unternehmensmodell. Springer
Wiendahl H-P, Breitenbach F, Klußmann JH, Nofen D (2005) Planung modularer Fabriken. Hanser, München
Wiendahl H-P, ElMaraghy HA, Nyhuis P, Zäh MF, Wiendahl H-H, Duffle N, Kolakowski M (2007) Changeable manufacturing—classification, design and operation. CIRP Ann Manuf Technol 56(2):783–809
Womack JP, Jones DT, Roos D (1990) The machine that changed the world. Rawson Associates, New York
Wooldridge M (2002) An introduction to multi-agent systems. Wiley
World Trade Organization (2013) World Trade Report 2013. http://www.wto.org/english/res_e/publications_e/wtr13_e.htm
World Urbanization Prospects: The 2014 Revision, Highlights, United Nations, Department of Economic and Social Affairs, Population Division (2014). https://esa.un.org/unpd/wup/publications/files/wup2014-highlights.Pdf. Accessed 11 Jan 2017
Zuehlke D (2010) Smartfactory-towards a factory-of-things. Annu Rev Control 34(1):129–138
Zukunft der Arbeit. Innovationen für die Arbeit von morgen, Bundesministerium für Bildung und Forschung, Bonn (2016). https://www.bmbf.de/pub/Zukunft_der_Arbeit.pdf. Accessed 11 Jan 2017

Chapter 2
Integrating Variable Renewable Electricity Supply into Manufacturing Systems

Jan Beier, Sebastian Thiede and Christoph Herrmann

Abstract Expanding renewable energy (RE) generation has been increasingly recognized as a central strategy for climate change mitigation. A substantial share of renewable energy generation comes from variable renewable energy sources (e.g. wind and solar), which are increasingly installed in decentralized structures. As such, integrating decentralized, variable RE generation into existing supply and demand structures is required to successfully further increase their share. Several different approaches and technologies are available for overcoming intertemporal and spatial demand and supply mismatches. Among them are conventional energy storage technologies as well as implicit energy storage options such as embodied energy in products, enabled through load shifting of energy-flexible production and manufacturing systems. This contribution begins with an overview of current challenges toward RE integration, followed by a discussion of available large-scale grid integration measures. Within the following, a focus is set on options for integrating decentralized variable RE. A promising approach is storing embodied energy in products. Its enabling method, energy flexibility of manufacturing systems, is detailed. A method to improve energy flexibility is discussed and its potential application demonstrated in a case study.

2.1 Decentralized Electricity Generation from Renewable Energy Sources

For decades, most energy systems have been relying on fossil fuel-based energy generation. Especially the energy and power sector is characterized by utilizing

---

J. Beier (✉)
The Boston Consulting Group, Cologne, Germany
e-mail: Beier.Jan@bcg.com

S. Thiede · C. Herrmann
Chair of Sustainable Manufacturing and Life Cycle Engineering, Institute of Machine Tools and Production Technology (IWF), Technical University Braunschweig, Braunschweig, Germany

S. Thiede and C. Herrmann (eds.), *Eco-Factories of the Future*, Sustainable Production, Life Cycle Engineering and Management,
https://doi.org/10.1007/978-3-319-93730-4_2

carbon-based fuels for electricity generation, among them black coal, lignite, natural gas, and petroleum. Global anthropogenic greenhouse gas (GHG) emissions from the energy supply sector amounted to approximately 35% of all GHG emissions in 2010. Without mitigation strategies, emissions from energy generation and utilization are expected to increase between 80 and 130% until 2050, compared to 2010. A central strategy for climate change mitigation is centered around switching energy and electricity generation from carbon-intensive (and thus GHG intensive) sources to alternative, less carbon-intensive sources such as renewable energy (RE) sources (IPCC 2014).

The importance of increasing the share of RE sources, especially for electricity generation, has been recognized by several nations, and thus targets for the increasing contribution of RE to (national) energy and electricity mixes have been formulated. For example, Germany has enacted a national policy called *Energiewende* (lit. *energy turnaround*) to reduce carbon emissions from energy and electricity generation while achieving a nuclear-free energy supply. Specifically, quantitative goals include cutting primary energy demand into half by 2050 compared to 2008 while increasing RE generation as share of final energy demand to 60%. Further, electricity generation from renewable sources is targeted to contribute 80% of the gross electricity demand in 2050 (Bundesministerium für Wirtschaft und Energie (BMWi) 2014).

Consequently, a set of incentives to increase electricity generation from RE sources has been enacted. Guaranteed grid feed-in tariffs for RE electricity generation were enacted, providing required financial predictability to promote investments in RE electricity generation capacities. Figure 2.1 illustrates German renewable electricity generation in 2000 and 2014. From 2000 to 2014, electricity generation from renewable sources increased from 6% to more than 27% in 2014. In particular, generation from wind and solar increased most significantly, with two-digit annual growth rates, surpassing all other renewable sources in 2014. Considering remaining opportunities for deployment of RE generation sources, (offshore) wind and additional photovoltaic (PV) systems for solar-based generation are expected to further increase their share toward fulfillment of stated RE energy generation goals within the next years (Bundesministerium für Wirtschaft und Energie (BMWi) 2014). Another factor is the relatively high social acceptance of RE generation, arising, for example, from environmental concerns, which supports future increase in RE penetration, also in other regions than Europe (Yuan et al. 2015).

When RE starts to supply a substantial share of total energy and electricity demand, integration into an existing energy system needs to be performed. Some RE sources are characterized by intermittent availability, i.e. power/electricity output varies over time. Output from conventional power generation facilities, e.g. fossil fuel or nuclear power plants, is controllable. Depending on the specific technology, electricity output can be increased or decreased, with varying time periods required for adjustment (e.g. minute ramp-up and -down times for gas turbines and up to multi-hour lead times for coal-fired plants). For the case of variable renewable energy (VRE)-based electricity generation, output can generally only be decreased. For example, wind and solar generation are, among others, dependent upon local weather conditions and solar radiation. Output can be reduced, e.g. by reducing the rotation speed of a

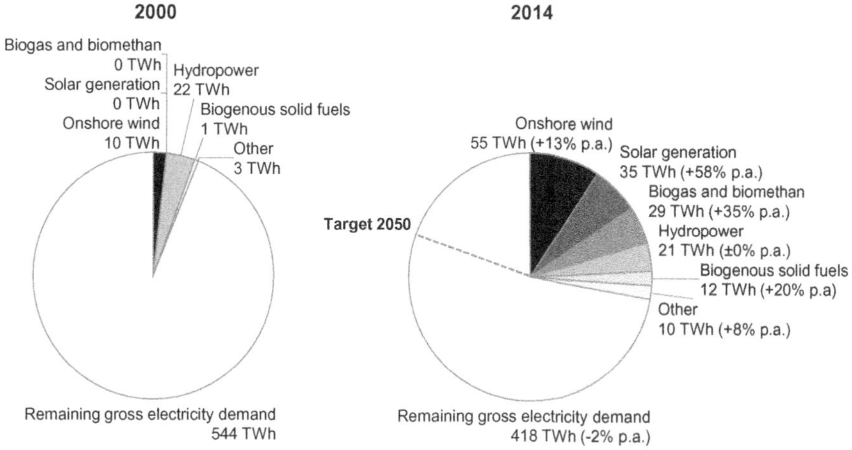

Fig. 2.1 German renewable electricity generation by source and remaining gross electricity demand in 2000 and 2014, percentage increase per annum (p.a.) from 2000 to 2014 (Bundesministerium für Wirtschaft und Energie (BMWi) 2015, own calculations)

wind turbine, while it cannot be increased (if output has not been curtailed before). Further, reducing output is usually unfavorable as marginal costs of VRE generation are low compared to other generation sources and maximum output is targeted to fulfill RE goals (IPCC 2012). To illustrate variation and time dependency, Fig. 2.2 shows wind and solar outputs from installations based in Braunschweig, Germany, over five days in October 2013. The first two days are characterized by high solar output during daytime and high, but fluctuating wind output, also during the night. The third day depicts a change in weather conditions. Solar output fluctuates during the day, indicating partly cloudy weather, while wind output declines at the end of the day. During the last illustrated two days, wind and solar outputs are both very low (10–20%) compared to the first three days. In summary, VRE generation can substantially vary over time, with second, minute, hourly, daily, and seasonal changes. Predicting VRE generation is a complex and difficult task due to inherent dynamics and stochastic behavior of several influencing factors. Substantial effort has been made to increase reliability of forecasts, for example, conditioning artificial neural networks (Qazi et al. 2015). However, residual deviations are unavoidable as a result of stochastic components (IPCC 2014).

Another characteristic of VRE is its decentralized generation. Wind and solar resources are usually geographically spread-out, i.e. geographic conditions determine the potential for RE generation (e.g. high wind resources are usually found within coastal regions, solar radiation increases in southern regions and with height). Consequently, electricity generation takes place where efficiency is expected to be high. Technologies which are used for decentralized generation are PV, wind-based generation, small hydropower, and biomass. Decentralized generation is expected to also increase within the next years (IPCC 2014).

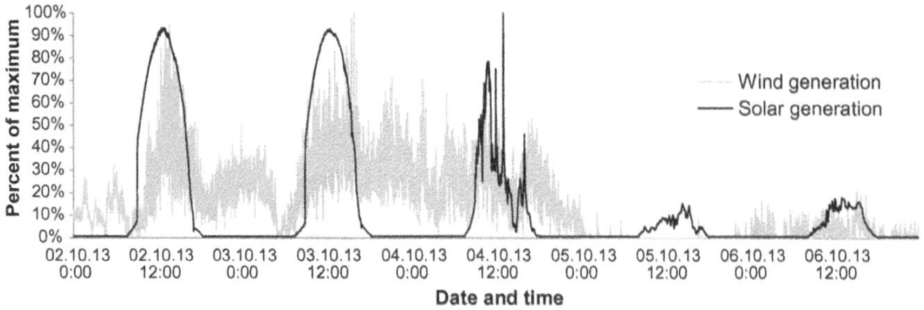

Fig. 2.2 Wind and solar generation at IWF, TU Braunschweig (latitude 52.2767, longitude 10.5369) from October 2 to 6, 2013, one-minute averages (second sample period), normalized to period's maximum demand

From an industry perspective, decentralized (on-site VRE) generation plays a vital role to reduce cost and increase independence from grid supply. About 18% of total industrial electricity demand was supplied by industry itself in Germany in 2013, which is equivalent to 45 TWh or 8.9% of Germany's gross electricity generation (DESTATIS 2014, 2015, own calculations). In 2013, approx. 16% of all companies have enacted measures for own supply in Germany, which is expected to increase in the following years. With 62%, realized and planned PV generation had the largest share of all utilized technologies (DIHK and VEA 2014). Due to economic benefits (e.g. reduced network charges), companies usually aim at maximizing the demand of on-site generated electricity.

Taking above considerations toward mitigating climate change through RE generation capacity increase, combined with decentralized generation in industry, into account, integrating VRE into an existing grid infrastructure is of special interest. As outlined above, VRE generation can only be influenced to a certain extent, and dispatchability is limited. Consequently, measures are required to overcome intertemporal and local discrepancies between volatile electricity supply and demand. Aligning electricity demand and supply can either be targeted on a grid level, i.e. large-scale storage, or performed on-site for the case of decentralized generation. Within the following, options for both strategies are discussed (compare to Fig. 2.3), starting with large-scale grid storage options.

2.2 Large-Scale Electricity Storage for Grid Integration

On a power grid level, large-scale electricity storage aims at overcoming temporal demand and supply mismatches, traditionally caused by low demand during night-time and a (relatively large) baseload power plant fleet with limited adjustability (e.g. nuclear power plants). Common energy storage technologies currently used on

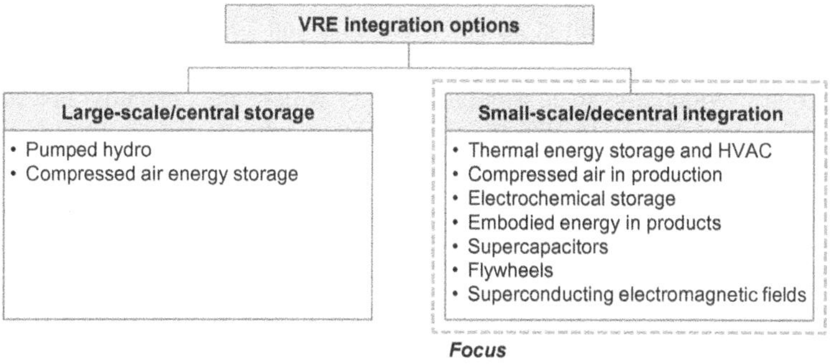

Fig. 2.3 Overview of electricity storage options

a commercial scale include pumped hydro storage and compressed air energy storage (CAES) (IPCC 2012).

Pumped storage is regarded as the only economic method for large-scale electricity storage at present (Deutsche Energie-Agentur GmbH (dena) 2010). During charging, water is pumped in a reservoir which lies geographically elevated. During discharging and thus electricity generation, the process is reversed, and water is released through turbines for electricity generation. However, remaining potential for additional pumped storage is, depending on the region, limited due to lacking geographic conditions, ecological concerns or political resistance (Germany Trade and Invest 2015). For the case of Central Europe, options for additional pumped storage in Scandinavian countries and the Alps are investigated, which would cause substantial transportation requirements. However, Norway, for example, has the potential for an additional 10–25 GW pumped storage capacity, which would result in about doubling the currently available capacity (IPCC 2012).

CAES plants store energy in compressed air, e.g. in an underground cavern. During excessive electricity supply periods, air from the surrounding environment is compressed into the cavern. However, during discharge and thus electricity generation, heat is required as the air temperature is reduced during expansion. Heat exchangers (storing heat from compression) or additional firing, e.g. using natural gas, are commonly used. Currently operated CAES plants include the 1978-built, 290 MW plant in Huntorf, Germany, and a 110 MW plant from 1991 in McIntosh, Alabama, USA (Crotogino et al. 2001). Initially, CAES was built to compensate mentioned demand/supply mismatches between night and day caused by baseload power plants. CAES integration into a system with a large VRE share has recently been investigated, among others due to their fast response times (Lund and Salgi 2009; Lund et al. 2009). Conversion inefficiencies of around 50% and related challenges for economically profitable operation currently prevent widespread application of CAES. Recent developments mainly target increasing conversion efficiency, e.g. adiabatic compressed air energy storage technology (AA-CAES) with an approximate conversion efficiency of 70% and, with an expected efficiency of 80%, isobaric adiabatic compressed air energy storage plant with combined cycle (ISACOAST-CC)

(Zunft et al. 2006; Bullough et al. 2004; Karellas and Tzouganatos 2014; Nielsen and Leithner 2009).

Other grid-wide/large-scale storage options which are currently investigated include batteries (chemical/redox-flow), fuel cells and thus hydrogen storage, (other) power to gas methods, flywheels, and supercapacitors (Deutsche Energie-Agentur GmbH (dena) 2010). However, technical (e.g. high energy losses in flywheels over time), environmental (e.g. batteries), and economic (e.g. conversion and economic efficiency of fuel cells) challenges prevent current implementation for large-scale electric energy storage. However, some of the mentioned technologies are used for providing ancillary services, e.g. frequency control.

2.3 Options for Integrating Decentralized Generation

Integrating decentralized generation into an existing power grid can either be targeted by applying previously discussed measures on a grid-wide scale or by increasing direct demand of on-site generated electricity. Increasing direct demand for decentralized electricity has multiple benefits: From a utility perspective, less transportation requirements and thus grid infrastructure is required to supply companies with electricity and to transport surplus on-site generated electricity to other customers. Further, if on-site electricity generation demand achieves to reduce volatility in grid demand and supply, less backup generation resources are required. From a company perspective, direct demand is favorable as companies which have on-site generation facilities usually face a lower marginal cost of on-site generation than grid supply costs, i.e. demanding a kWh of on-site generated electricity is usually cheaper than demanding one kWh from the public grid. Related to this are additional economic incentives, for example, avoiding grid charges and taxes (e.g. subsidies for RE integration, for example, under the German *Erneuerbare Energien Gesetz*) (DIHK and VEA 2014).

Within the following, options for increasing direct demand of on-site VRE generation are discussed. A focus is set on a production/manufacturing perspective. For each option, the general idea is briefly discussed, followed by advantages and disadvantages. Challenges toward implementation and main influences, which need to be considered, are included.

2.3.1 Thermal Energy Storage and HVAC

Thermal energy storage can be used for VRE generation in two different ways: (1) thermal storage can be used to store heat, e.g. from solar radiation, to be converted to electricity at a later point in time and (2) electricity can be converted to thermal energy for later use, e.g. by converting thermal energy back to electricity or for heating/cooling purposes (which might substitute electricity use) (Denholm et al. 2010).

Utilizing thermal storage for later conversion to electricity is, for example, applied in concentrated solar power plants. However, round-trip efficiency for electricity-heat-electricity cycles is usually relatively low.

Utilizing thermal energy storage for air conditioning (cooling/heating) purposes is a well-known approach to accommodate variable energy supply and/or to profit from variable prices. Storage options include thermal building mass, where a building's heat capacity is used for pre-cooling or -heating in periods where high (VRE) electricity supply is present (Braun 2003; Reynders et al. 2013). Utilizing building thermal mass to shift electricity demand for heating is considered a promising approach toward influencing the demand for VRE integration (Reynders et al. 2013). Further, chilled water and ice, e.g. on rooftop tanks or submerged underground, can be used to increase control range and/or heat capacity (Ashok and Banerjee 2003; Arteconi et al. 2012; Rankin and Rousseau 2008). This might also include utilizing underground rock formations for heat/cool storage capacities, where hot/cold water is pumped into suitable geological patterns, which can even be used for seasonal load shifting.

Related to cooling and heating applications as part of heating, ventilation, and air conditioning (HVAC) is ventilation, especially in industrial setting. Clean air is used for load shifting, i.e. ventilation is switched-off during low electricity availability if emission levels allow for reduced ventilation, and emission reduction is performed during periods of high electricity availability. Further, increased heating might be required if ventilation is switched on, which can increase emission-related load-shifting opportunities (Junge 2007).

Other thermal storage applications, especially in the context of VRE integration into a production facility, include process cooling and heating. For example, supercritical steam can be stored and supply process heat in production facilities (Oertel 2008). Further technologies include thermo oil and solid materials, while the latter is not (yet) commercially established. Process cooling can be supplied similarly, e.g. by chilled water or ice storage.

From a technology perspective, thermal energy storage can be realized by changing the temperature of a material or by latent thermal storage (Oertel 2008; Arteconi et al. 2012). The latter principle is based on a change of state of matter, i.e. from liquid to solid. These so-called phase-change materials have the advantage of storing a significantly higher amount of energy (per volume/mass), while energy flows are realized at nearly constant temperature. However, challenges include low heat transfer capabilities, requiring large surfaces for transmission.

2.3.2 Compressed Air

Compressed air (CA) used in production and manufacturing can be used as direct energy storage option. In general, compressed air can be generated during times of high electricity availability and stored in tanks and within the CA system volume (e.g. pipes). Shifting CA production can, for example, be used to profit from fluctuating electricity prices (Kleiser and Rauth 2013) or to integrate VRE generation into a production/manufacturing facility (Beier et al. 2015). An advantage compared to e.g. CAES or battery storage is the avoidance of an additional conversion cycle, as CA is generated and stored for direct use in production. CA is required independent of the point in time when it is generated. Therefore, conversion inefficiencies are not a result of possible load-shifting and related energy conversion actions, as long as technical parameters are held constant. Nonetheless, in order to increase the amount of storable energy, the operating pressure band needs to be adjusted and/or the CA system (tank) size increased. While the first measure can increase compression inefficiency, the latter requires additional investments. Further, CA losses need to be accounted for, which can also increase as a result of load-shifting actions (e.g. higher pressure level over longer periods of time increases losses through leaks). Another limitation is provided by production system characteristics: only the share of total energy demand which is supplied by CA can be used for load shifting, i.e. even with high compressor capacities and large storage tanks, total CA demanded by processes and facilities defines the amount of load that can be potentially shifted. In addition, operational constraints for compressors, such as switch-on time and maximum allowed number of switches to prevent motor wear-out need to be taken into consideration (Ruppelt 2003).

2.3.3 Electrochemical Storage

Considered electrochemical storage options for VRE integration include conventional, high-temperature and redox-flow batteries. Conventional rechargeable batteries include, for example, lead–acid batteries, nickel–cadmium, sodium–sulfur, lithium-ion, zinc–air and lithium-air batteries (Oberhofer 2012; Deutsche Energie-Agentur GmbH (dena) 2010). Under the goal of VRE integration, most relevant parameters include energy and power density (e.g. per mass and volume), investment/operating costs and environmental impact (caused mainly by utilized materials). As the amount of different battery technologies indicates, a wide range of technical and economic parameters can be achieved. For example, lead–acid batteries are a mature technology which is frequently used for hybrid energy systems, e.g. VRE and battery/diesel backup generation (Bernal-Agustín and Dufo-López 2009). Their relatively low energy density is offset by possible high discharging power (which has also established their use as starter batteries in cars) and relatively low cost. However, availability of other, more economic storage options in a grid infrastructure (such as pumped hydro) has so far limited lead–acid battery applications mostly to stand-alone/island scenarios where no grid access is provided (Oberhofer 2012). A second battery type is lithium-based (e.g. lithium-ion, lithium polymer, lithium–air) batteries. They currently undergo a rapid development toward their use in mobile applications such as electric vehicles (Gallagher and Nelson 2014). Their energy density compared to lead–acid batteries is relatively high, cycle lifetime is significantly increasing as a result of ongoing research efforts and the main material, lithium (and currently graphite) is an abundant mineral resource (Oberhofer 2012). However, current technical parameters combined with costs make lithium-based batteries mainly suitable for balancing power applications in a VRE context.

Considering advantages and disadvantages of batteries for VRE integration, charging and discharging efficiency of batteries can be very high with values above 95%. Self-discharge losses are relatively low, with single-digit percentages per month (depending on technology and operating parameters such as temperature) (Deutsche Energie-Agentur GmbH (dena) 2010). Response rates are fast, making them suitable for both decentralized energy storage and balancing power-related applications. Depending on technology, batteries can impose serious environmental and operational hazards, which can include toxicity (e.g. large amounts of lead in lead–acid batteries) or flammability when ruptured (e.g., lithium is flammable when exposed to atmospheric moisture) (Oberhofer 2012). Further, especially in a VRE setting where demand and supply peaks are shaved, batteries might experience a high number of charge and discharge cycles, causing fast deterioration and wear-out, reducing economic and ecologic efficiency (Deutsche Energie-Agentur GmbH (dena) 2010). However, current research and development efforts are indicating that especially lithium-ion batteries might be a promising technology for future VRE integration.

High-temperature batteries, among them the mature sodium–sulfur (NaS) and the newer sodium–nickel chloride (ZEBRA) battery, have a solid electrolyte and liquid electrodes (Oertel 2008; Oberhofer 2012). As such, temperatures above 300 °C are required to keep the electrodes molten. Challenges include operational safety (fires) and reduced efficiency due to required heating. However, their energy density is relatively high and their lifespan is favorable. For example, Japan extensively uses NaS batteries, mostly for balancing power. Recent developments indicate that operational and economic improvements can make high-temperature batteries suitable for VRE integration.

The technical principle of redox-flow batteries is similar to conventional batteries, with the difference that the electrolyte within a cell can be exchanged during operation. This feature allows advanced configuration of redox-flow batteries, i.e. power and energy can be changed separately from each other. Capacity can be adjusted by altering the size of electrolyte storage tanks. Recharging the battery is possible by replacing the electrolyte, which favors mobile applications, e.g. in electric vehicles. Cycle and temporal lifetime, start-up time and efficiency (around 80%) are relatively high. However, costs prohibit large-scale industrial applications if other (e.g. compressed air, pumped storage) options are available. Common technologies include vanadium redox and zinc–bromine redox-flow batteries (Denholm et al. 2010; Deutsche Energie-Agentur GmbH (dena) 2010; Oberhofer 2012; Oertel 2008).

In summary, electrochemical storage options are currently mainly used for hybrid-/stand-alone energy systems and to provide balancing and regulatory power. Cycle lifetime, costs and environmental concerns are main barriers for increased utilization for VRE integration. However, strong R&D efforts cause electrochemical storage to become increasingly competitive.

2.3.4 Embodied Energy in Products

Embodied energy in products can be used to store energy by shifting production from times with low electricity availability to times with high electricity availability. As such, embodied energy in products can contribute to integrate VRE and/or account for volatile supply and/or prices in general. In contrast to other storage options, recovering energy is not targeted, i.e. the stored energy is already in the form of its desired end-use. For example, energy can be stored in chemical bounds, e.g. in free aluminum during aluminum electrolysis (reduction of aluminum oxide), steelmaking from iron ore or, in general, contained in chemicals obtained by (endothermic) reactions (Allwood and Cullen 2012). Another example is joining two workpieces, e.g. by drilling, threading, and connecting both parts with a screw. Most of the energy demanded during the process is dissipated into the environment as a result of inefficiencies and thus energy storage is rather latent. However, assuming that a value-creating process requires a certain amount of energy to be fulfilled, this total energy can be regarded as embodied energy in the product (for an overview of energy demand modeling of manufacturing processes and discussion of components see, e.g., Gutowski et al. 2006). Consequently, utilizing, for example, VRE during value creation of a product and not (fossil fuel)-based supply improves integration of VRE and reduces demand for conventional supply. Nonetheless, opportunities for increasing demand of VRE and thus load shifting are subject to operational constraints, most notably potential impact on throughput and technical constraints. Further, opportunities to increase demand of VRE supply is limited to energy/electricity demand of processes. For example, a battery could be used to supply electricity to processes and, e.g. HVAC facilities and thus increase the use of VRE for both applications. Embodied energy flow into products is limited to the influenceable demand of processes. Further, the amount of embodied energy which can be stored is strongly dependent on (intermediate) product storage opportunities (more storage space results in more embodied energy storage), i.e. energy demand of processes can only be flexibly adjusted if material flow is sufficiently decoupled.

The inherent advantage of embodied energy storage is avoidance of additional energy conversion (similar to compressed air). Assuming that electricity demand is only shifted and not decreased or increased in its total amount, and holding (technical) parameters constant, results in no additional inefficiencies (compared to traditional energy storage technologies such as pumped hydro, CAES or batteries). However, the dynamic nature of production and manufacturing systems, compared with nonlinear energy demand (e.g. idle and production demand), requires detailed evaluation of embodied energy storage opportunities and their change of operational, technological, and environmental impacts.

From a production technology perspective, embodied energy in products to implicitly store (renewable) energy is discussed in the context of energy flexibility of manufacturing systems (while other options to flexibly adjust energy demand, such as battery storage, might also be included in a wider energy flexibility definition). Energy flexibility can be defined as "[...] the ability of a production system to adapt itself fast and without remarkable costs to changes in energy markets." (Graßl et al. 2013). Examples which explicitly consider intermediate product storage to shift electricity demand of manufacturing systems include (Ashok 2006; Middelberg et al. 2009; Li et al. 2012; Fernandez et al. 2013; Sun et al. 2014; Schultz et al. 2015; Keller et al. 2015; Beier et al. 2015). Utilizing energy flexibility of manufacturing systems to integrate VRE, especially decentralized generation, is regarded as a promising approach and can reduce requirements for traditional energy storage, which involve additional conversion cycles and infrastructure. Further, utilizing energy flexibility of industry provides additional opportunities for creating new business models (Lorenz et al. 2012). Dependent upon specific system design and technology and thus required investments and additional operating costs, companies can generate additional profit by, e.g. reduced energy costs and participating in reserve markets.

Energy flexibility of manufacturing systems can also be used for grid-wide integration of VRE. From a power system operator's/utility's perspective, demand-side management (DSM) encompasses the two main strategies energy efficiency and demand response (Paulus and Borggrefe 2011). Demand response aims at reshaping demand of customers to increase demand and supply matching, but is, in contrast to energy flexibility, driven by utilities (and not an electricity customer-centered method) (Elkarmi and AbuShikhah 2012). Common strategies include price-based and incentive-based programs, i.e. customers are faced with time-dependent electricity prices or they receive an incentive if demand is altered upon request from the utility (Albadi and El-Saadany 2008).

2.3.5 Supercapacitors, Flywheels, Superconducting Electromagnetic Fields

Supercapacitors (also called ultracapacitors), flywheels, and superconducting electromagnetic fields (sometimes called superconducting magnetic energy storage (SMES)) are summarized as they are rather used for power quality and frequency regulation than for substantial (VRE-) energy storage (Denholm et al. 2010).

Supercapacitors are mostly used for frequency regulation and to smooth output/stabilize power in VRE-based electricity generation settings (Seo et al. 2010; Denholm et al. 2010). Supercapacitors can reach an energy density beyond 20 kWh per cubic meter, and they exhibit high cycle stability values with more than 100,000 charge and discharge cycles. Their efficiency might reach 80–95%, and maintenance costs are relatively low compared to other available technologies. Costs were estimated to be around 320 EUR per kW, while both energy density and cost are expected to improve significantly in the future (Deutsche Energie-Agentur GmbH (dena) 2010). However, due to their relatively low energy capacity, supercapacitors are unlikely to be used for extended energy storage, also in the near future.

Flywheels store energy in a rotation mass by converting electricity into kinetic energy during charging and back to electricity during discharging. Fast response but limited capacity make them especially useful for frequency regulation rather than (long-term) energy storage (Denholm et al. 2010). Single flywheels can have capacities of 100 kWh and are installed in larger arrays to increase capacity. Their relatively high cycle efficiency of more than 90% is offset by high discharge rates of several percents per hour (Deutsche Energie-Agentur GmbH (dena) 2010). Advantages of flywheels include their relatively high reliability and utilization of less hazardous materials than, for example, in batteries. Further, expected lifespan and cycle stability are high, with nearly no maintenance required if magnetic bearings are used (Oberhofer 2012). One of the first large-scale flywheel energy storage plants started operation in 2011 in Stephentown, New York (Beacon Power 2015). The plant has a capacity of 5 MWh at 20 MW charge and discharge rate and consequently mainly participates in reserve markets. However, current large-scale application still faces economic challenges, indicated by, e.g. bankruptcy of the first operator of the described plant. Therefore, utilizing flywheels for decentralized VRE integration is only suitable under specific circumstances, e.g. high temperatures or if hazardous materials cannot be used.

SMES can reach efficiency values of beyond 90% with very fast response times, which is offset by their relatively low capability to store energy (Oberhofer 2012; Denholm et al. 2010). As such, they are mostly used for power quality and frequency stabilization. In addition, costly setup and operation, partly induced by required cooling to achieve superconductivity, combined with two-percent discharge losses per day, prevent SMES to be (solely) applied for VRE integration.

Storage technologies discussed in this subsection (supercapacitors, flywheels, SMES) are also used in hybrid storage technologies to increase efficiency. For example, one application is their combination with conventional batteries in electric vehicles to improve efficiency and response to short-term, high-power requests, e.g. during acceleration or braking (energy recovery) (Neugebauer 2014; Butterbach et al. 2011).

2.3.6 *Intermediate Summary*

Several different approaches and technologies exist to directly and indirectly store energy from decentralized VRE generation for integration into industry and manufacturing systems. An overview, including advantages and disadvantages and factors for application, can be found in Table 2.1. Depending on system structure, VRE supply and individual goals for VRE generation and integration, different options are available. These options need to be evaluated in detail, e.g. under economic, ecological, and operational aspects, to support decision making.

In order to integrate decentralized VRE-based electricity generation, embodied energy in products, compressed air and battery storage, including electric vehicles, are promising approaches. As mentioned, energy storage in embodied energy of products and flexible compressed air generation has the advantage that no additional conversion cycle is required (as, for example, compared to battery storage). However,

Table 2.1 Selected energy storage options for decentralized VRE integration into production and manufacturing systems

Storage technology	General principle	Advantages	Disadvantages	Factors for application
Thermal energy storage and HVAC	Store energy in heat/cooling capacities (temperature and phase-changing) and emission levels	• Very high capacity • Direct use can be cost efficient (heating/cooling) • Depending on technology, losses relatively low • Safe and environmentally friendly	• Conversion efficiency (to electricity) low • Recovering electricity from thermal energy potentially uneconomic • Requires available thermal mass • Limited to amount of required heating/cooling/ emission reduction if not converted back to electricity • Potential complex dynamics	Suitable for heating, ventilation and air conditioning and, depending on technical processes, to supply process cooling and heating when dynamics can be controlled
Compressed air (CA) in production	Shift CA production to times of high electricity availability, store energy in CA	• Infrastructure already present • Avoidance of additional conversion cycles • Fast response time • Environmentally friendly	• Limited by total CA demand • Increased inefficiency if technical parameters changed • Additional investment and cost if new installations required	Suitable where CA is required for production and CA storage capacities are sufficient
Electrochemical storage	Energy storage via chemical reactions	• Storage and supply of electricity (flexible use) • Several configurations possible • Fast response times • Potential for high energy storage	• (Not yet) cost efficient • High wear-out (limited cycles) • Environmental concerns	Depending on technology, used for peak shaving, as reserve capacity and energy storage: to be compared to other available options
Embodied energy in products	Store energy in embodied energy of products by shifting production demand	• No additional conversion cycle required • Utilization of existing infrastructure • Environmentally friendly	• Requires flexibility of production system • Limited to amount of flexible demand from machines/system • Might adversely affect technical, operational and other indicators	Applicable for production (systems) with sufficient energy flexibility: scheduling flexibility, product storage and direct control required
Supercapacitors	Energy storage in electric field	• Very fast response • Nearly no maintenance • High cycle stability • Moderate to high efficiency	• Relatively costly • Low volumetric energy density	Only suitable for frequency regulation/power quality application and power smoothing and/or very short-term storage
Flywheels	Energy storage in rotating mass	• Fast response • Environmentally friendly • Long lifespan • Low maintenance	• High self-discharge • Energy storage limited • High initial costs	For fast response, high temperature, environmental sensitive environment, short-term storage
Superconducting Magnetic Energy Storage	Energy storage in electro-magnetic field	• Very fast response • High cycle efficiency • Low environmental impact and hazardous materials	• Lower efficiency during stand-by (cooling) • Costly to set-up and operate • Very low energy density • Discharge losses	Useful for frequency regulation, power quality applications (e.g. uninterrupted supply)

as a result of inherent dynamics of energy supply and manufacturing systems, several challenges exist. Within the following, energy flexibility of manufacturing systems to enable embodied energy storage is discussed in more detail.

2.4 Enabling, Improving and Evaluating Embodied Energy Storage in Manufacturing Systems

Energy flexibility of manufacturing systems for embodied VRE energy storage is enabled through load shifting of manufacturing system energy demand. Within this section, a focus is set on electricity demand as a primary energy carrier in most manufacturing systems and on embodied energy storage using energy flexibility as an emerging method to integrate VRE. However, discussed approaches, including challenges and requirements toward implementation, are also applicable for other energy carriers, e.g. direct heat demand from solar radiation.

Figure 2.4 illustrates an example for load-shifting opportunities to accommodate (on-site, renewable) variable electricity generation (note that the given example is for illustrative purposes only and includes no measured values). Before load shifting, demand is characterized by two (flattened) peaks, e.g. from running production equipment with a dedicated production and thus energy demand cycle. The two peaks are followed by a fluctuating demand pattern. Electricity supply exhibits an unstable pattern with small-scale, short-term volatility combined with longer term, larger volatility, for example, found in wind generation output patterns (compare also to Fig. 2.2). The difference between supply and demand is compensated by a connected grid, i.e. if demand is higher than on-site supply, electricity is drawn from the grid and, for the case of an excessive on-site generation, electricity is fed into the grid. Before load shifting, total absolute difference between demand and supply (area under the curve from the lower graph in Fig. 2.4) is 399 kW · time units (203 kW · time units grid supply and 196 kW · time units grid feed). Shifting demand by postponing the two peaks in demand and scheduling the fluctuating demand first results in an improved match between demand and supply (right graphs of Fig. 2.4). Total absolute deviation is decreased to 171 kW · time units (89 kW · time units grid supply and 82 kW · time units grid feed), resulting in more than halving grid support requirements (electricity supply and feed-in). However, load-shifting opportunities depend on process characteristics, system dynamics, and operational constraints, e.g. if the fluctuating pattern has been caused by cleaning or setting-up the machine *after* processing two parts which caused the spikes, load shifting is not possible (at least not in the way the example assumes).

The capability of a manufacturing system to store embodied energy in products via load shifting is influenced by process-related and system-related factors. For example, Graßl defines a dimensionless indicator for evaluating energy flexibility of a manufacturing process (Graßl 2015). Among others, flexibility is enabled through the variation potential of process energy demand. For example, a machine which can

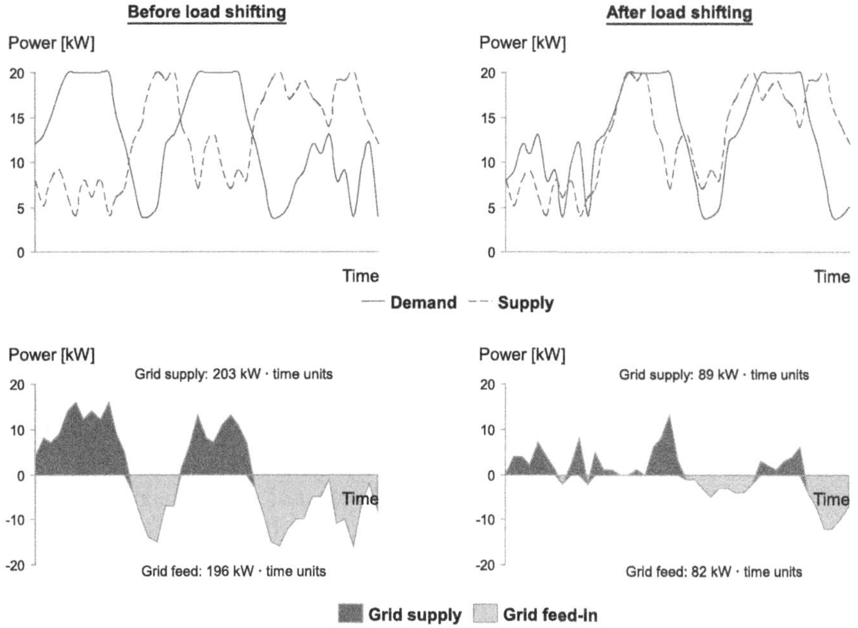

Fig. 2.4 Example for industrial load shifting under variable (renewable) electricity supply (illustrative example)

c.p. assume more different energy states than another machine is considered more energy flexible. Further, reflection of a broad energy demand continuum (i.e. different states with significantly different electricity demand as opposed to states with similar electricity demand) is beneficial for energy flexibility. In addition, further indicators such as costs or required time for adjustment are discussed. In (Graßl and Reinhart 2014), a set of measures to enable energy flexibility is discussed, among them altering machine schedules and process starts (compare to graphical load shifting example in Fig. 2.4), operational measures such as adjusting workforce schedules (free time and shift times) and switching of energy carriers.

For the case of interlinked manufacturing systems, discussed load-shifting methods for single processes are also applicable. However, material flow dynamics deserve special consideration as energy flexibility measures might have an impact on the overall system performance. In connected manufacturing systems (i.e. the number of possible ways a product can take through the system is fixed and product storage between processes is limited), rescheduling of a process can impact adjacent processes (upstream and downstream). For example, deferring a process start can block an upstream process if the storage capacity between the two processes is exhausted. For the case of downstream processes, processes might be starved if no intermediate products in a connecting buffer are available (Beier et al. 2015; Fernandez et al. 2013). For multi-product systems, altering schedules including job sequencing might cause a required input product type for downstream processes not to be available on time. From a total system perspective, rescheduling processes can result in missing required delivery time of products. Further, maximum throughput of a manufacturing system is determined by one or more bottleneck processes (Zhai et al. 2011). If a bottleneck process is rescheduled, blocked or starved (i.e. increasing nonproductive time), the total throughput of the system is reduced. As a consequence, energy flexibility benefits (such as integrating VRE and reducing grid electricity costs) need to be compared to economic disadvantages due to lost production (e.g. lost product margin of sales). Another option is introducing a control method which ensures that no throughput due to energy flexibility measures is lost.

In general, energy flexibility via embodied energy storage of a manufacturing system, i.e. how much energy can be stored in embodied energy, is influenced by two components: the amount of electricity demand that can be shifted and used for energy flexibility purposes and secondly the amount of products and thus embodied energy that can be stored. The first component is closely related to energy flexibility of single machines as outlined above: a machine which has a nearly fixed energy demand (e.g. a washing machine) independent of state/throughput (excluding switch-off) is not suitable for load-shifting purposes. In general, machines with a high idle electricity demand and small production-related demand only offer limited opportunities for rescheduling as their main demand (idling) cannot be influenced by production scheduling. Another aspect is the adjustment time of processes, i.e. a process which requires significant time to change its energy demand state compared to the frequency in which energy supply changes is less suitable for energy flexibility purposes than a process which can rapidly adapt itself. For example, a machining process with a cycle time of several minutes, which is not interruptible, might not be suitable to follow minute or second-based changes of electricity supply. However, combining inert processes and systems with short-term electricity storage (e.g. batteries, flywheels, capacitors) can overcome differences due to delayed adjustment or supply/demand peaks. The second important metric, embodied energy storage amount, is influenced by (intermediate) product storage opportunities. If more products can be stored in between processes, more opportunities (in general) exist to shift demand of a single process without impacting adjacent processes. For example, if processes can run through times of high electricity availability (e.g. a sunny day and PV generation) and store products in intermediate buffers, processes can be switched-off or set to idle longer during times of low electricity availability while adjacent (bottleneck) processes are not required to curtail production as a result of missing inputs or insufficient output product storage. Another influencing parameter is the position of product storage buffers within the value chain: installing storage after processes with low electricity demand and thus less added embodied energy compared to processes with high electricity demand and added embodied energy results in less embodied energy storage capabilities (compared to installation after a process with high energy demand and flexibility).

In general, energy flexibility of manufacturing systems and embodied energy storage are enabled and increased by multiple factors, including system design and operation. Further, energy flexibility and embodied energy storage in products can be enhanced by including, for example, flexible compressed air generation. However, mutual influences and dependencies of material and energy flows require a dynamic approach to enhance energy flexibility of manufacturing systems. The next section describes an initial system design and real-time control approach toward energy flexibility improvement and illustrates its applicability and effectiveness in relation to other energy storage options using an example case study.

2.5 Example Case Study: Self-Sufficient Manufacturing System

A case study is used to demonstrate the potential of integrating VRE into an (existing) manufacturing system via different (direct and indirect) energy storage options.

2.5.1 Concept Description

Figure 2.5 illustrates the basic steps of the proposed concept and their application in the example case study. Initially, energy flexibility objectives (1) need to be defined. Within this example, achieving an electricity self-sufficient factory/manufacturing system (i.e. autarkical factory), which is independent from external grid supply, is set as a goal. Based on this goal, hypotheses are formulated (2) to achieve the stated goal. For example, embodied energy storage, battery storage, energy efficiency measures, VRE supply capacity, and their combination could be investigated. In a third step, the considered dynamic system needs to be modeled (3), bearing in mind the formulated hypotheses. Dynamic system modeling is required to investigate dependencies between energy and material flows in a (highly nonlinear) system. The case study uses an example manufacturing system from an experimental laboratory in combination with recorded wind and solar supply data. The next step is to derive scenarios (4), which are capable of providing results for previously defined hypotheses. For example, different intermediate product storage sizes or battery sizes are examined. Dynamic system behavior is calculated based on model and scenario input parameters (5). In this study, a simulation prototype is used to evaluate the system behavior over time. Based on the obtained results for different indicators (e.g. self-sufficiency of the manufacturing system's electricity demand), results are interpreted and conclusions are drawn (6). If desired results toward defined objectives are achieved (7), the preferred solution can be implemented. If goals are not (yet) reached or additional findings result in new hypotheses to be tested, hypotheses are restated and re-tested. The next section describes the application and results in more detail.

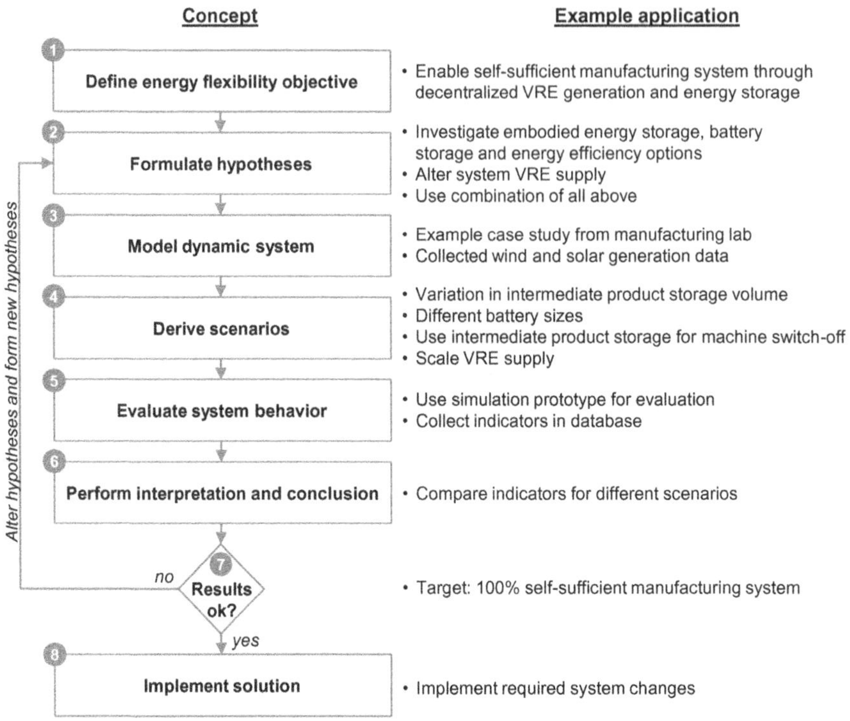

Fig. 2.5 Energy flexibility concept and example application case

2.5.2 *Case Study Manufacturing System and Results*

The considered manufacturing system layout and energy demand parameters are illustrated in Fig. 2.6. The manufacturing process that consists of nine connected steps, cycle time, and state-based energy demand data has been obtained. The process *CNC* (last process) has the longest cycle time of all processes (72 s) and thus establishes the bottleneck of the system. To enable some operational flexibility, the manufacturing system is assumed to have a target production rate of one product every 80 s, thus denoting 90% system output of maximum possible output. A compressor park is attached to the system with three compressors and a seven cubic meters storage tank. Energy supply profile data has been collected from wind and solar generation facilities located in Braunschweig, Germany, from September 3 to 31, 2013, with one-second sample rate and aggregated to one-minute arithmetic average values.

In order to achieve an energy self-sufficient manufacturing system, different hypotheses are formulated (compare to Fig. 2.5) and scenarios derived. In particular, the following experiments are conducted:

	Distri-bution A	Distri-bution B	Trans-port A	Assem-bling	Press	Switch	Furnace	Trans-port B	CNC
Cycle time	8 sec.	2.5 sec.	4 sec.	10 sec.	8 sec.	6 sec.	17 sec.	4.5 sec.	72 sec.
Elec. idle	960 W	990 W	1,000 W	1,400 W	920 W	1,160 W	4,430 W	1,080 W	4,060 W
Elec. prod.	3,230 W	3,150 W	3,120 W	1,950 W	1,080 W	2,500 W	6,070 W	2,860 W	7,800 W
CA idle		380 W	140 W	140 W	620 W	110 W		120 W	120 W
CA prod.		1,000 W	400 W	1,970 W	2,590 W	120 W		320 W	140 W

H

Process Buffer

Fig. 2.6 Case study manufacturing system process structure and parameters (scaled by factor 100 from original system)

- **Buffer size** indicating different intermediate product storage capacity is varied from 5 to 700 pieces in every buffer between processes. In order to enable energy flexibility, an energy-related scheduling is required. The scheduling approach described in Beier et al. (2015) is applied, which finds the combination of non-throughput critical processes for which system demand is expected to fit VRE supply best, considering operating constraints such as non-interruptible processing. Both manufacturing processes and compressors are controlled according to energy flexibility targets (for additional background on, e.g. CA tank size variation to support flexible CA generation see also Beier et al. (2015)).
- **Battery size** is varied from 0 to 700 kWh capacity, with charge and discharge time of three hours in every scenario, and a one-piece flow for products (replacement of a product if withdrawn from a process' outgoing buffer, compare also to Beier et al. (2015)).
- **Different VRE supply levels** indicated by the ratio between total VRE supply and total system demand under one-piece flow control over total simulation time are varied. For example, a VRE factor of 1.2 indicates that average VRE supply is 1.2 higher than average total system electricity demand.
- **Process switch-off** indicates that processes switch themselves off if a one-minute idle wait time has passed and sufficient products are in outgoing buffers to have downstream processes prevented from being starved (which is also an energy efficiency control measure). For the case of battery size variation and process switch-off, an intermediate product storage of 200 pieces has been chosen, with switched-on energy flexibility control.

The results for described scenarios can be found in Fig. 2.7. The indicator *self-sufficiency* is used to describe the system's time-dependent energy demand ratio from VRE supply (i.e. a 70% self-sufficiency indicates that 70% of total electricity demand was supplied by VRE, considering temporal dynamics). Constant throughput was achieved for all scenarios.

Looking into different intermediate product storage levels (left graph in Fig. 2.7), self-sufficiency can be increased by up to four percent by deploying additional buffer

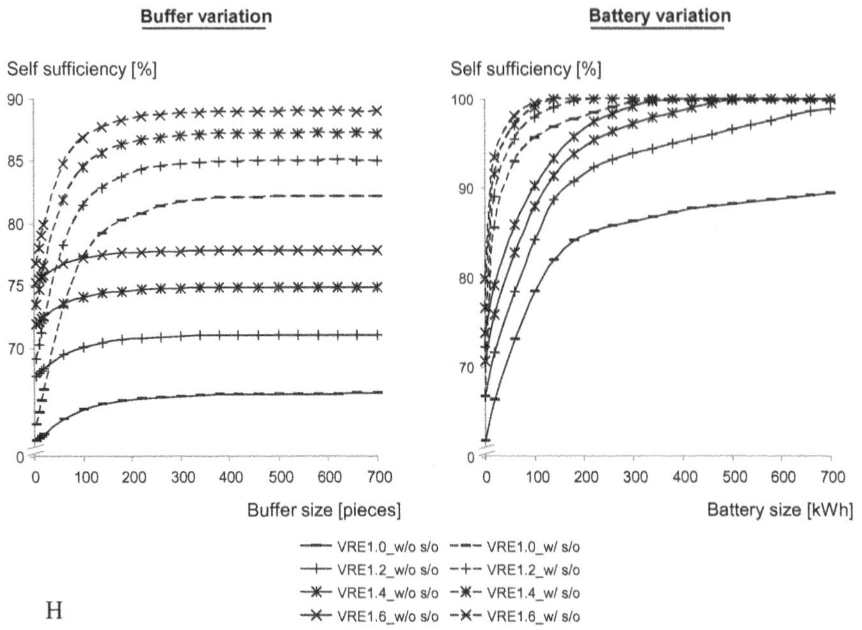

Fig. 2.7 Values for self-sufficiency indicator under different scenarios (w/o s/o: without process switch-off, w/ s/o: with process switch-off)

capacities (for a 1.0 VRE supply case) compared to a case with low buffer capacity. However, increased intermediate product storage is only beneficial up to a certain level (between 200 and 300 pieces), where marginal improvement is getting very low. Further, intermediate product storage is relatively more beneficial for lower VRE supply scenarios than for high VRE scenarios. In summary, the example illustrates that the amount of embodied energy storage is defined by both flexible and thus variable demand of processes and intermediate product storage. Increasing product storage above a certain level is not beneficial, as the variable share of demand limits further increase in embodied energy storage potential. For the case of switching processes off when enough intermediate products are available (dashed lines in left graph), self-sufficiency can be significantly increased if enough intermediate product storage is available. However, even with high VRE supply levels and process switch-off, increasing self-sufficiency beyond 90% is not achievable.

The right graph of Fig. 2.7 illustrates the results for different battery capacity values. First, considering different VRE supply levels without embodied energy storage, battery capacity beyond 700 kWh is required to enable full self-sufficiency for the case of 1.0 and 1.2 VRE supply capacity. The first full self-sufficient case within evaluated scenarios is achieved at approx. 500 kWh battery storage and 1.4 times VRE supply. However, including energy flexibility and thus added embodied energy storage and process switch-off, a fully autarkical scenario can be obtained with a 350 kWh battery for a 1.0 VRE supply ratio, and a 150 kWh battery is sufficient in a 1.6 VRE supply case. Further, significant improvement, also below full autarkical operation, can be achieved with small batteries in combination with energy flexibility.

In summary, the case study illustrates that self-sufficiency and thus energy storage and integration of VRE can be enabled by different methods, including embodied energy storage and battery storage. Embodied energy storage can result in higher utilization of VRE by increasing intermediate product storage. Nonetheless, in order to obtain a fully self-sufficient setup, battery storage is unavoidable. However, required battery storage can be significantly reduced in combination with embodied energy storage. A combination of different storage options, including, for example, embodied energy storage and batteries, appears to be a favorable solution. Depending on system-specific values, substantial economic and environmental benefits might be gained by avoiding additional battery capacity or avoiding VRE overcapacity. However, operational indicators such as system residence time and increased inventory levels might need further detailing before deriving a (cost efficient) strategy.

2.6 Discussion, Conclusion, and Outlook

Integrating VRE into an existing electricity grid and decentralized generation into a local grid can be achieved utilizing different methods to match volatile energy supply and demand. Most common methods include electricity storage through conversion into an energy form/carrier which is easily storable, e.g. potential energy. For example, pumped hydro storage is the most common application for large-scale electricity storage. For the case of decentralized VRE generation, especially wind and solar-based electricity generation, currently available energy storage technologies include, for example, thermal energy storage or electrochemical batteries. However, all technologies which involve additional conversion cycles induce inefficiencies, i.e. the amount of usable energy is reduced. A novel approach is embodied energy storage, for example, in intermediate products or compressed air. The difference is the avoidance of an additional conversion cycle, i.e. rather than storing VRE in a battery through charging and discharging, manufacturing systems can be enabled to process products or generate CA when VRE is available and reduce energy demand when VRE is not available. This method is enabled by decoupling manufacturing processes, i.e. by introducing intermediate product storage or CA storage. Within this context, the term energy flexibility can be used to describe the ability of a manufacturing system to adapt energy demand to volatile supply. However, several challenges exist, among them operational challenges such as keeping throughput constant or avoiding excessive increase of products' system residence time. As such, a manufacturing system control approach is required to enable energy flexibility of manufacturing systems.

A concept for successfully integrating and improving energy flexibility of manufacturing systems is presented. A case study from a manufacturing laboratory is used to test different hypotheses toward full self-sufficiency of the manufacturing line. The results indicate that a combination of embodied energy storage and battery storage is a promising approach to enable electricity autarky of the system.

Summarized, new methods such as embodied energy storage can complement traditional energy storage options for VRE integration. Their clear advantage is the avoidance of (substantial) additional infrastructure (e.g. batteries) and conversion inefficiencies. However, operational challenges and dynamic system behavior require careful evaluation of energy flexibility methods. A combination of available options, e.g. thermal storage, batteries, embodied energy storage, and potentially dispatchable backup generation (e.g. diesel generator) can be used to successfully integrate decentralized VRE generation. Further, grid feed-in and central, large-scale storage can complement decentralized integration efforts. Available measures can support each other and partly be substituted for each other. As such, a preferred/optimal solution for widespread VRE deployment and integration is likely to include several options in combination. Multi-criteria evaluation and combined modeling of methods are required to find this preferred solution, which is a central research lead in the context of VRE deployment to mitigate climate change.

References

Albadi MH, El-Saadany EF (2008) A summary of demand response in electricity markets. Electr Power Syst Res 78:1989–1996

Allwood JM, Cullen JM (2012) Sustainable materials: with both eyes open. UIT, Cambridge

Arteconi A, Hewitt N, Polonara F (2012) State of the art of thermal storage for demand-side management. Appl Energy 93:371–389

Ashok S (2006) Peak-load management in steel plants. Appl Energy 83(5):413–424

Ashok S, Banerjee R (2003) Optimal cool storage capacity for load management. Energy 28(2):115–126

Beacon Power (2015). http://beaconpower.com/. Accessed 13 Nov 2015

Beier J, Thiede S, Herrmann C (2015) Increasing energy flexibility of manufacturing systems through flexible compressed air generation. Procedia CIRP 37:18–23

Bernal-Agustín JL, Dufo-López R (2009) Simulation and optimization of stand-alone hybrid renewable energy systems. Renew Sustain Energy Rev 13(8):2111–2118

Braun JE (2003) Load control using building thermal mass. J Sol Energy Eng 125(3):292–301

Bullough C, Gatzen C, Jakiel C, Koller M, Nowi A, Zunft S (2004) Advanced adiabatic compressed air energy storage for the integration of wind energy. In: Proceedings of European wind energy conference EWEC 2004, London, pp 22–25

Bundesministerium für Wirtschaft und Energie (BMWi) (2014) Zweiter Monitoring-Bericht "Energie der Zukunft". Technical report, Berlin

Bundesministerium für Wirtschaft und Energie (BMWi) (2015) Erneuerbare Energien - Zeitreihen Erneuerbare Energien. http://www.erneuerbare-energien.de/EE/Navigation/DE/Service/Erneuerbare_Energien_in_Zahlen/Zeitreihen/zeitreihen.html. Accessed 14 Dec 2015

Butterbach S, Vulturescu B, Forgez C, Coquery G, Friedrich G (2011) Lead-acid battery model for hybrid energy storage. In: 2011 IEEE vehicle power and propulsion conference, Chicago, IL. IEEE, pp 1–5

Crotogino F, Mohmeyer K-U, Scharf R (2001) Huntorf CAES: more than 20 years of successful operation. Solut Min Res Inst Spring Meet. Orlando, FL, pp 351–357

Denholm P, Ela E, Kirby B, Milligan M (2010) The role of energy storage with renewable electricity generation. Technical report, National Renewable Energy Laboratory, Golden, CO. http://www.nrel.gov/docs/fy10osti/47187.pdf

DESTATIS (2014) Industriebetriebe produzieren knapp 9% der in Deutschland erzeugten Strommenge. https://www.destatis.de/DE/PresseService/Presse/Pressemitteilungen/2014/10/PD14_363_433.html. Accessed 16 Mar 2015

DESTATIS (2015) Energieverwendung der Betriebe im Verarbeitenden Gewerbe 2013. https://www.destatis.de/DE/ZahlenFakten/Wirtschaftsbereiche/Energie/Verwendung/Tabellen/KohleErdgasStrom.html. Accessed 16 Mar 2015

Deutsche Energie-Agentur GmbH (dena) (2010) Dena grid study II. Integration of renewable energy sources in the German power supply system from 2015–2020 with an outlook to 2025. Technical report, Berlin

DIHK and VEA (2014) Faktenpapier Eigenerzeugung von Strom: Rahmenbedingungen, Trends. Beispiele, Technical report, Berlin, Brüssel, Hannover

Elkarmi F, AbuShikhah N (2012) Power system planning technologies and applications: concepts. solutions and management, engineering science reference, Hershey, PA

Fernandez M, Li L, Sun Z (2013) "Just-for-Peak" buffer inventory for peak electricity demand reduction of manufacturing systems. Int J Prod Econ 146(1):178–184

Gallagher KG, Nelson PA (2014) Manufacturing costs of batteries for electric vehicles. In: Pistoia G (ed) Lithium-ion batteries: advances and applications. Elsevier B.V, Amsterdam et al, pp 97–126

Germany trade and invest (2015) The energy storage market in Germany. Technical report, Köln

Graßl M (2015) Bewertung der Energieflexibilität in der Produktion. Herbert Utz Verlag, München

Graßl M, Reinhart G (2014) Evaluating measures for adapting the energy demand of a production system to volatile energy prices. Procedia CIRP 15:129–134

Graßl M, Vikdahl E, Reinhart G (2013) A petri-net based approach for evaluating energy flexibility of production machines. In: Zäh M (ed) Proceedings of the 5th international conference on changeable, agile, reconfigurable and virtual production (CARV 2013). Springer, Switzerland, Munich, pp 303–308

Gutowski TG, Dahmus J, Thiriez A (2006) Electrical energy requirements for manufacturing processes. In: 13th CIRP international conference on life cycle engineering, Lueven, pp 623–627

IPCC (2012) Renewable energy sources and climate change mitigation: special report of the Intergovernmental Panel on Climate Change. Technical report, Intergovernmental Panel on Climate Change, Cambridge. https://doi.org/10.5860/CHOICE.49-6309

IPCC (2014) Climate change 2014: mitigation of climate change. working group III contribution to the IPCC 5th assessment report. Chapter 7: energy systems. In: Climatic change (2014) Mitigation of climate change working group III contribution to IPCC 5th assessment report. Cambridge University Press, Cambridge, New York, NY

Junge M (2007) Simulationsgestützte Entwicklung und Optimierung einer energieeffizienten Produktionssteuerung. kassel university press GmbH, Kassel

Karellas S, Tzouganatos N (2014) Comparison of the performance of compressed-air and hydrogen energy storage systems: Karpathos island case study. Renew Sustain Energy Rev 29:865–882

Keller F, Schönborn C, Reinhart G (2015) Energy-orientated machine scheduling for hybrid flow shops. Procedia CIRP 29:156–161

Kleiser G, Rauth V (2013) Dynamic modelling of compressed air energy storage for small-scale industry applications. Int J Energy Eng 3(3):127–137

Li L, Sun Z, Tang Z (2012) Real time electricity demand response for sustainable manufacturing systems: challenges and a case study. In: IEEE international conference on automation science and engineering, Seoul, pp 353–357. http://ieeexplore.ieee.org/lpdocs/epic03/wrapper.htm?arnumber=6386437

Lorenz S, Putz M, Schlegel A (2012) Energieeffizienz 2.0: Neue Geschäftsmodelle auch für die Industrie. ZWF Zeitschrift für wirtschaftlichen Fabrikbetr 107(9):599–602

Lund H, Salgi G (2009) The role of compressed air energy storage (CAES) in future sustainable energy systems. Energy Convers Manag 50(5):1172–1179

Lund H, Salgi G, Elmegaard B, Andersen AN (2009) Optimal operation strategies of compressed air energy storage (CAES) on electricity spot markets with fluctuating prices. Appl Therm Eng 29(5–6):799–806

Middelberg A, Zhang J, Xia X (2009) An optimal control model for load shifting—with application in the energy management of a colliery. Appl Energy 86:1266–1273

Neugebauer R (2014) Handbuch Ressourcenorientierte Produktion. Carl Hanser Verlag GmbH und Co, KG, Wien

Nielsen L, Leithner R (2009) Dynamic simulation of an innovative compressed air energy storage plant—detailed modelling of the storage cavern. WSEAS Trans Power Syst 4(8):253–263

Oberhofer A (2012) Energy storage technologies and their role in renewable integration. Technical report, Global Energy Network Institute. https://www.geni.org/globalenergy/research/energy-storage-technologies/Energy-Storage-Technologies.pdf

Oertel D (2008) Energiespeicher - Stand und Perspektiven: Sachstandsbericht zum Monitoring "Nachhaltige Energieversorgung". Technical report, TAB - Büro für Technikfolgen-Abschätzung beim Deutschen Bundestag, Berlin

Paulus M, Borggrefe F (2011) The potential of demand-side management in energy-intensive industries for electricity markets in Germany. Appl Energy 88(2):432–441

Qazi A, Fayaz H, Wadi A, Raj RG, Rahim N (2015) The artificial neural network for solar radiation prediction and designing solar systems: a systematic literature review. J Clean Prod 104:1–12

Rankin R, Rousseau P (2008) Demand side management in South Africa at industrial residence water heating systems using in line water heating methodology. Energy Convers Manag 49(1):62–74

Reynders G, Nuytten T, Saelens D (2013) Potential of structural thermal mass for demand-side management in dwellings. Build Environ 64:187–199

Ruppel E (2003) Druckluft Handbuch. 4th edn, Essen

Schultz C, Sellmaier P, Reinhart G (2015) An approach for energy-oriented production control using energy flexibility. Procedia CIRP 29:197–202

Seo H-R, Kim G-H, Kim S-Y, Kim N, Lee H-G, Hwang C, Park M, Yu I-K (2010) Power quality control strategy for grid-connected renewable energy sources using PV array and supercapacitor. In: International conference on electric power systems 2010. Incheon, Korea, pp 437–441

Sun Z, Li L, Fernandez M, Wang J (2014) Inventory control for peak electricity demand reduction of manufacturing systems considering the tradeoff between production loss and energy savings. J Clean Prod 82:84–93

Yuan X, Zuo J, Huisingh D (2015) Social acceptance of wind power: a case study of Shandong Province, China. J Clean Prod 92:168–178

Zhai Y, Sun S, Wang J, Niu G (2011) Job shop bottleneck detection based on orthogonal experiment. Comput Ind Eng 61(3):872–880

Zunft S, Jakiel C, Koller M, Bullough C (2006) Adiabatic compressed air energy storage for the grid integration of wind power. In: Sixth international workshop on large-scale integration of wind power and transmission networks for offshore windfarms, Delft, pp 1–6

Chapter 3
Development of a Sustainability Assessment Tool for Manufacturing Companies

Nadine Madanchi, Sebastian Thiede, Manbir Sohdi and Christoph Herrmann

Abstract Regarding the increasing global population and the related demand for natural resources, the societal and political demand for a sustainable development has increased significantly over the last decades. Although the traditional definition of a sustainable development "development that meets the needs of the present without compromising the ability of future generations to meet their own needs" (WCED, Our Common Future (the Brundtland Report), Oxford, 1987) appears comprehensible, it still poses a challenge to actually assess a sustainable development. As a result, sustainability assessment is becoming a rapidly developing area with a growing number of frameworks and tools with a wide range of different focus levels. Despite the variety, many of these tools are not adaptable for manufacturing companies. They are, for example, too general or focus only on specific elements. Furthermore, the existing tools usually require a lot of effort and insight data in order to be applied. Against this background, the development of a tool at the level of manufacturing companies is presented. Based on existing integrated sustainability assessment tools, a set of indicators is compiled and integrated into a framework that calculates an overall composite index combining different indicators. The developed tool distinguishes itself from other tools because it can be used from an external as well as from an internal perspective and it allows the assessment of a manufacturing company's overall and relative sustainability with minimal time effort.

N. Madanchi (✉) · S. Thiede · C. Herrmann
Sustainable Manufacturing & Life Cycle Engineering Research Group,
Institute of Machine Tools and Production Technology (IWF),
Technical University Braunschweig, Braunschweig, Germany
e-mail: n.madanchi@tu-bs.de

M. Sohdi
School of Engineering, University of Rhode Island, Kingston, USA

S. Thiede and C. Herrmann (eds.), *Eco-Factories of the Future*, Sustainable Production, Life Cycle Engineering and Management,
https://doi.org/10.1007/978-3-319-93730-4_3

3.1 Introduction

For the last two centuries, industry and economy have evolved on the premise that the earth is an unlimited "store of resources" and a stable ecosystem (Graedel and Allenby 2010). However, as the population exceeds 7 billion and the standards of living improve enormously, the attention and awareness toward the limited natural resources increase as well. The general interest should be to use resources consciously in order to satisfy human demand (Davidson et al. 2010). One approach to this challenge can be found in the key concept of sustainability. By regarding the three dimensions—society, environment, and economy—it aims for our society to meet present as well as future needs worldwide. Obviously, manufacturing is a major factor in this approach toward a more sustainable society (Gutowski et al. 2013). Against this background, many manufacturing companies have already started to reconsider the idea of being "green" and how to deal with sustainability. However, this change of attitude was of course supported by even more factors. Environmental regulations have a significant impact, especially if they are supported by legal consequences. Therefore, many global manufacturers feared to be locked out of the market if they do not change their policies toward the concept of sustainability (Srinivasan 2011). Furthermore, investors are also interested in the sustainability performance of manufacturing companies and an increasing number integrates it into their portfolio decisions. They are one of the target groups that use indexes and tools to evaluate manufacturing companies. This trend toward socially responsible investing is another important factor that forced companies to adapt their strategy (DJSI 2013).

Although there has been a lot of work on researching sustainability on different levels, the research on rapid sustainability assessment tools at the level of manufacturing companies and factories is still insufficient. None of the existing tools fulfills the criteria of a generic applicability and holistic view on sustainability simultaneously (Chen et al. 2013). Thus, the paper focuses on this gap and additionally aims at the external perspective of a company without insight information. For this purpose, a tool is being developed that can be used as decision support by benchmarking unknown manufacturing companies and by evaluating them over time.

A brief overview of different state-of-the-art sustainability assessment tools is presented and discussed in Sect. 3.2 of this chapter. In the following section, the methodological procedure on the development of the sustainability assessment tool is presented, along with the selected key performance indicators and the framework of the tool. Section 3.4 applies the tool in a case study and presents the results.

3.2 Literature Review—Categorization of Sustainability Assessment Tools and Indicators

Over the last years, a number of researchers and organizations focused on the challenge not only how to become sustainable but also how to assess it. They have defined frameworks and developed tools that can be used to assess sustainability of different subjects in different ways and have proved effective in practice. In the literature, several authors categorized these tools and frameworks based on numerous factors and dimensions. For example, Ness et al. conducted an overview of tools by considering the focus of the tool (i.e., product level or policy), the temporal characteristics, and the degree to which it integrates environmental, social, and/or economic aspects (Ness et al. 2007). Feng et al. on the other hand categorized sustainable assessment tools into a hierarchy of global, country, sector, corporation, process, and product levels (Feng et al. 2010a, b). Labuschagne et al. conducted an overview of tools that include a set of indicators, integrate all three dimensions of sustainability, have a wide focus, and are independent (Labuschagne et al. 2005). Additionally, Chen et al. evaluated representative tools that are most related to manufacturing companies by four criteria: generic applicability, rapid assessment, application on factory level, and holistic view of sustainability (Chen et al. 2013).

To demonstrate the research background, this paper combines the different classifications and categorizes tools by considering the following three factors:

- Integration of all three dimensions of sustainability, i.e., if the tool considers environmental, social, and economic aspects.
- Hierarchy/focus level, i.e., if the focus is on the global, country, sector, corporation, or product level.
- Scope of application, i.e., if the tool was developed by a company or by an organization.

The developed categorization and overview of sustainability assessment tools are illustrated in Fig. 3.1 including the most quoted tools at each level. It consists of two main branches: the non-integrated and the integrated indicators. The non-integrated indicators include indicators that do not consider all three dimensions of sustainability simultaneously, but they are especially designed for one dimension and are widely used. Therefore, they are further broken down into development-based, economy-based, and ecosystem-based indices. The second branch on the other hand covers all integrated tools and divides them first into macro- and micro-tools and subsequently into a hierarchy of global, country, sector, corporate, and product level. While the macro-tools are developed by superordinate organizations, the micro-tools are developed by a company. This separation is based on the main issue of macro-frameworks and macro-tools. Their focus is mainly "on the external reporting for stakeholders, rather than on internal information need to decision-making and re-design or optimization for actual eco-innovation" (Feng and Joung 2009, p. 2). The tools developed by a company (micro-tools) on the other hand give the manufacturers the possibility to evaluate and track the sustainability performance within their own environment. However, the issue with those tools concerns the fact that they are designed mainly for the specific environment of a company or supply chain. Therefore, it is important to include both and distinguish them in a comprehensive overview. The distinction between intra- and inter-company is additionally important, because it shows whether the tool is used internally within one company or between companies, for example, to assess suppliers.

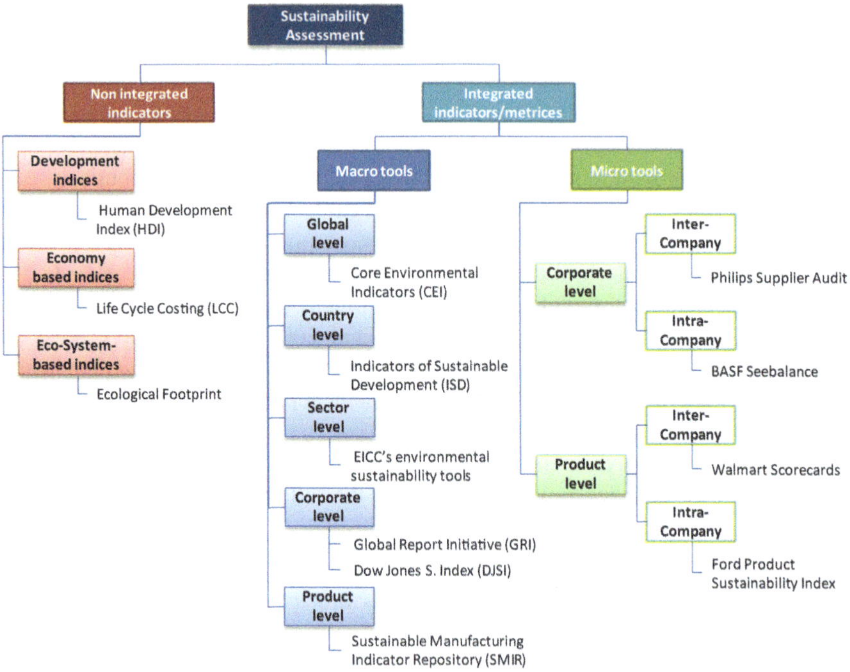

Fig. 3.1 Categorization of assessment tools

The categorization of common sustainability assessment tools demonstrates that a wide range of tools with a focus on different subjects exists in practice. However, each category of tools shows different characteristics and a different area of application. The tools categorized as micro-tools have proven to be designed specifically for their industry and do not apply for general use. Additionally, they require a lot of insight data, because they are intended to be used internally within a company or supply chain. Regarding the tools categorized as macro-tools, it shows that most of them are not related to manufacturing companies. However, the application of the tools that are actually related to manufacturing companies usually requires a great quantity of data and effort in order to be applied.

Based on these results, this paper aims to develop a tool that assesses a manufacturing company's sustainability performance and ensures a rapid and integrated assessment for all industries with a minimal time effort. Its main characteristic, however, concerns the data the tool is based on. It should be possible to use the tool as an external user without internal information, meaning the data for the indicators should be available through published sustainability reports, Web pages, etc. Therefore, it should be possible to evaluate two or more unknown companies and assess them as alternatives against each other. It should also be possible to evaluate one company over time, in order to observe its sustainable development. Against this background, the tool's purpose is intended for external investors as main users who integrate sustainability consideration into their portfolio. The tool is supposed to provide a quick and general overview of the sustainability performance and shall support the comparison of different alternatives. At the same time, internal factory managers may also use the tool to compare themselves to other companies or to identify possible improvements or deteriorations in terms of sustainability.

The overview of existing sustainability assessment tools provides the basis upon which the new assessment tool will be created in the following of this paper.

3.3 Methodology

With regard to the described purpose of the new assessment tool, this chapter presents the methodology for developing it. The first step presents general criteria for indicators and a hierarchy of the framework. Based on those criteria, appropriate themes and indicators are then derived for each dimension. The derived indicators are further judged as to whether an indicator has a positive or negative influence on a company's sustainability. The next step requires a normalization of the indicators to avoid adding up incompatible data sets that can lead to inaccuracies in further steps. After evaluating and normalizing the indicators, they also have to be weighted in order to obtain a meaningful evaluation of the sustainability performance. From this model, it is possible to calculate a subindex for each sustainability dimension. Finally, all three subindices are combined into one overall composite sustainable performance index. The process is visualized in Fig. 3.2.

3.3.1 Criteria of Sustainability Performance Indicators

Indicators provide key information about a condition or state in a numerical form on the basis of a scale. Thus, indicators are usually a step beyond primary data (Veleva and Ellenbecker 2001). They vary depending on the type of system they monitor. In terms of this study, sustainability indicators can be defined as "information used to measure and motivate progress toward sustainable goals" (Ranganathan 1988, p. 2). However, there are certain characteristics that effective sustainability indicators have in common. The Sustainable Measures Group (Sustainable Measures 2010) as well as Anderson et al. and Feng et al. (Feng et al. 2010a, b) have established the following criteria:

- *Measurable*: Indicators need to be capable of being measured quantitatively or qualitatively.
- *Relevant*: Indicators have to fit the purpose of measuring sustainability performance and provide useful information on it.
- *Understandable*: Indicators should be easily understood by people who are not experts.

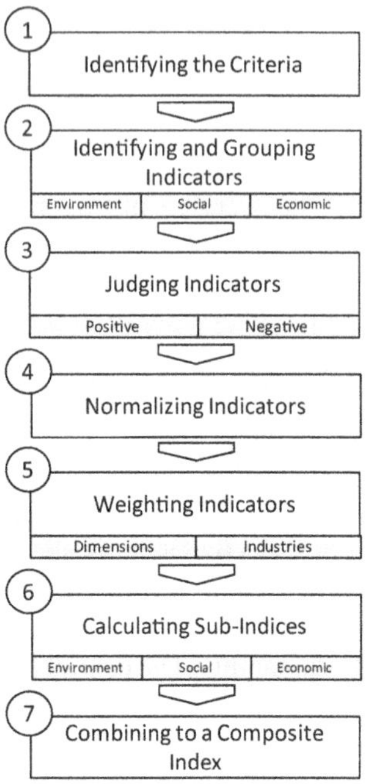

Fig. 3.2 Methodology for developing a framework to assess factory sustainability

- *Manageable*: Indicators have to be limited to the minimal number required to meet the purpose of measuring.
- *Reliable*: Indicators need to provide trustworthy information.
- *Data accessible*: Indicators have to be based on information that are available or can be easily accessed.
- *Timely manner*: Indicators should be measured on a regular basis to enable timely, informative decision-making.

The characteristics listed above help to distinguish indicators from primary data, goals, parameters, or issues.

3.3.2 *Identifying and Grouping of Sustainability Performance Indicators*

In order to identify and group indicators, it is necessary to define a hierarchical structure for the framework (see Fig. 3.3).

According to this hierarchical structure, the tool consists of different superordinate and general dimensions, which are further divided into themes. These themes describe a broader subject area within the dimensions and include several indicators. As shown in the figure, the dimensions are derived from the literature. This can be done easily because the general literature focuses on the three traditional dimen-

Fig. 3.3 Structure of the framework for sustainability assessment

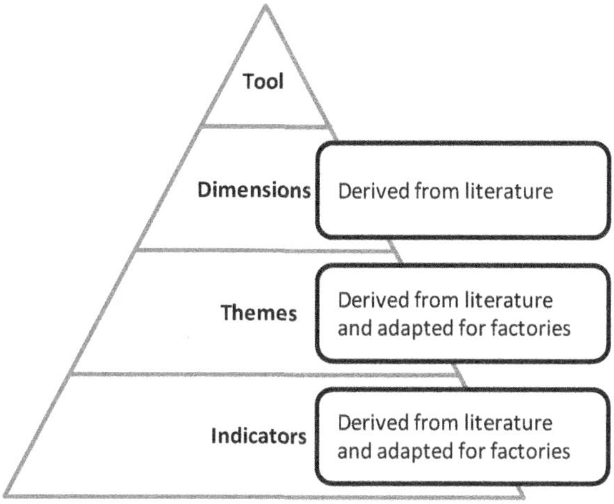

sions of sustainability: social, environment, and economy. Therefore, the framework adopts this view and contains the same three dimensions. In contrast to the dimensions, the themes and indicators require more effort as each sustainability tool in the literature focuses on different aspects. Therefore, it is important to analyze and compare the main sustainability assessment tools that have already been identified in the literature review above, with regard to the relevance to the assessment of manufacturing companies. Table 3.1 organizes the most important sustainability tools from the previous chapter by focus level, dimension, themes, and subthemes. Based on this information, it is possible to derive dimension-specific themes, which are discussed in the following section.

3.3.2.1 Themes for Environmental Sustainability

The environmental dimension traditionally gains most of the attention in terms of sustainability, and it is the dimension discussed in most detail in the literature. Therefore, current-integrated tools use a wide range of themes to evaluate the environmental performance. With regard to Table 3.1, however, it can be identified that most tools use relatively similar themes and subthemes. Furthermore, all of them focus on the external impact on the environmental system. Based on Table 3.1, the following main themes are derived:

(a) Natural resources and assets: This theme assesses a company's use of energy, water, and material as well as the amount of waste created by the manufacturing company.
(b) Pollution: This theme evaluates a factory's contribution to climate change and global warming. Additionally, it takes substances into account that present hazards to human health or the environment. This theme is commonly expressed by the global warming potential and acidification potential, which are not inde-

Table 3.1 Overview of existing frameworks and derivation of a new framework

Dim	Theme/ subtheme	Global	Country	Sector		Corporate		Pro	Duct	Factory
		CEI-OECD	UN-CSD-ISD	EICC	GRI	DJSI	BASF	FPSI	Walmart	New tool
Environmental	**Pollution issues**	**x**	**x**	**x**	**x**	**x**	**x**	**x**	**x**	**x**
	Emission, effluent	x	x	x	x		x	x	x	x
	Climate change	x	x			x				
	Toxicity potential	x					x			x
	Permits and reporting			x		x				
	restricted/ hazardous									
	Substances			x				x		
	Risk potential			x			x			
	Natural resources and assets	**x**	**x**	**x**	**x**	**x**	**x**	**x**	**x**	**x**
	Solid wastes	x		x	x		x			x
	Water	x	x	x	x	x	x			x
	Energy	x		x	x	x	x		x	x
	Material	x		x			x	x	x	x
	Biodiversity	x	x		x	x				
	Land use	x	x				x			
	Oceans, seas, and coasts	x	x							
	Compliance				x					
	Natural hazards		x							

(continued)

Table 3.1 (continued)

Dim	Theme/ subtheme	Global	Country	Sector		Corporate		Pro	Duct	Factory
		CEI-OECD	UN-CSD-ISD	EICC	GRI	DJSI	BASF	FPSI	Walmart	New tool
Social	**Health and safety**		**x**	**x**	**x**	**x**	**x**	**x**	**x**	**x**
	For employees		x	x	x	x	x			x
	For customers				x		x	x		
	Working accidents			x	x		x			x
	Machine safeguarding			x						
	Industrial hygiene and									
	Toxicity potential			x			x			x
	Physically demanding work			x						
	Emergency preparedness			x						
	Mortality		x							
	Sanitation, food, and housing		x	x						
	Labor practices and development		**x**	**x**	**x**	**x**	**x**			**x**
	Training/ education		x	x	x		x			x
	Satisfaction (strikes)						x			

(continued)

Table 3.1 (continued)

Dim	Theme/ subtheme	Global	Country	Sector		Corporate		Pro	Duct	Factory
		CEI-OECD	UN-CSD-ISD	EICC	GRI	DJSI	BASF	FPSI	Walmart	New tool
	Wages and benefits			x			x			x
	Working hours			x						
	Human rights and decent work			**x**	**x**		**x**			**x**
	Non-discrimination			x	x					
	Freedom of association			x	x					
	child labor avoidance			x	x					
	Freely chosen employment			x	x					
	Gender equality				x		x			x
	Integration of handicapped people				x		x			
	Part-time workers						x			
	Governance and community		**x**		**x**	**x**	**x**			
	Corruption		x		x					
	Security/crime		x				x			

(continued)

Table 3.1 (continued)

Dim	Theme/ subtheme	Global	Country	Sector		Corporate		Pro	Duct	Factory
		CEI-OECD	UN-CSD-ISD	EICC	GRI	DJSI	BASF	FPSI	Walmart	New tool
	Investment					x	x			
	Public policy				x					
	Demographics	**x**	**x**							
	Population change	x	x							
Economic	**Management**			x		x				
	Brand management					x				
	Risk and crisis management			x		x				
	Stakeholders engagement					x				
	Performance and development	**x**	**x**		**x**	**x**	**x**		**x**	**x**
	Innovation, R&D					x			x	x
	Market presence				x					
	Indirect economic Impacts	x			x					
	Exports/trade	x	x				x			
	Financials						x	x		x
	Material costs						x			
	Energy costs						x			
	Profit margins						x			x

pendent from other themes and indicators, such as the use of energy or material, but they are widely used in practice and can often be found in public records.

3.3.2.2 Themes for Social Sustainability

Recently, the public and especially stakeholders have expanded the focus from environmental-related to social-related issues. Therefore, businesses pay increasingly more attention to the social dimension of sustainability, although the work on this topic is still insufficient (Labuschagne et al. 2005). It is striking that the more modern tools like Electronic Industry Code of Conduct (EICC) and BASF SEEBALANCE contain significantly more social aspects than the older tools, such as Dow Jones Sustainability Index (DJSI) or Core Environmental Indicators of the Organization for Economic Cooperation and Development (CEI-OECD).

In contrast to the environmental dimension, most of the tools considering the social dimension have an internal view (influence on the social system of the company) instead of an external view (influence of the company on the social system of the broader society). Since the tool developed within this paper is aimed at assessing the social sustainability at the manufacturing company, the focus is also internal. The following themes are derived from Table 3.1 and describe the main issues of the social dimension with regard to manufacturing companies:

(c) Health and safety: This theme focuses on the security and well-being of all employees. It evaluates the preventative measures as well as the risk potential.
(d) Labor development and work satisfaction: This theme assesses the general working conditions and the continuous development of the employees and their talents.
(e) Equal opportunity and decent work: This theme evaluates the compliance of equal rights and fair employment practice standards. It contains aspects such as gender equality and equal career chances.

3.3.2.3 Themes for Economic Sustainability

In terms of economic sustainability, the review of current-integrated frameworks from Sect. 3.2 shows that there are two different understandings of economic sustainability. Since OECD and United Nations Commission on Sustainable Development (UN-CSD) are located at the global and national level, it is obvious that they take impacts from the economic system at the national and global levels into account. However, Global Reporting Initiative (GRI) assesses sustainability at a company level and considers "organization's impacts on the economic circumstances of its stakeholders and on economic systems at the local, national, and global levels" (Global Reporting Initiative 2011). All three frameworks focus on the general economic performance and development (see Table 3.1). However, the DJSI and the EICC consider economic performance in terms of the internal management, while the BASF SEEBALANCE and Ford Product Sustainability Index (FPSI) frameworks attempt to minimize their costs (see Table 3.1). Thus, there are two different approaches to choose from: One approach takes the external impacts on the entire economic systems into consideration, while the other approach focuses on the internal economic impacts of a business. With regard to the statement that the first goal of businesses toward sustainability is to stay in business, the focus within this study is internal. Activities at the factory level contribute to the overall profitability of the company and only subsequently contribute to the economic system on a broader, national level (Labuschagne et al. 2005). Therefore, the following themes are derived based on the DJSI, EICC, BASF, FPSI, and Walmart scorecard:

(f) Financials: This theme takes the internal financial stability of factories into account by assessing the profits.
(g) Development: This theme focuses on the investment and expenditures on future development and environment, health, and safety compliance.

3.3.2.4 Sustainability Performance Indicators

After defining themes for each dimension and the general criteria for indicators, it is now required to define and constrain the concept to a number of key performance indicators that meet all the criteria and can be measured, monitored, and recorded on a regular basis. A wide range of possible sustainability performance indicators can be found in the literature (see Sect. 3.2). However, every indicator is not relevant to industry and can be evaluated from an external perspective. Therefore, suitable key indicators have to be identified, by comparing the existing tools from Sect. 3.2 and identifying the most common key indicators. Additional sets of indicators found in the literature that focus on sustainable manufacturing are included as well: Krajnc and Glavič (2005), and Veleva and Ellenbecker (2001). These indicators are then tested and compared with sustainability reports published by different companies to ensure the data availability for the external use of the tool. To achieve this, the BMW Group Sustainable Value Report 2012, the BASF Report 2012, and the AkzoNobel Report 2012 are analyzed. These reports are published annually by the companies to report their figures and goals in terms of sustainability.

Generally, the study aims for using only quantitative indicators, as these are more objective and less biased than qualitative ones. It should also be possible to express each indicator in relative terms and not only in absolute terms, as different manufacturing companies have to be compared on a meaningful level. Social indicators, for example, should be expressed relative to the size of the workforce and environmental indicators relative to an appropriate measure of production, such as produced units of product or an indication of produced weight. The identified key sustainability indicators along with their dimensions, themes, and units are summarized in Table 3.2.

Table 3.2 Key performance indicators of factory sustainability

Dimension	Theme	Indicator	Unit
Environment	(a) Natural resources and assets	1. Energy use	MWh
		2. Material use	kg
		3. Freshwater consumption	m^3
		4. Waste generation	kg
	(b) Pollution	5. Global warming potential	t CO_2-eq
		6. Acidification potential	t SO_2-eq
Social	(c) Health and safety	7. Working accidents	–
		8. Safety training	–
		9. Hazardous materials	kg
	(d) Labor development and work satisfaction	10. Training and education	h
		11. Sickness frequency	Days
		12. Employee attrition rate	%
	(e) Equal opportunity and decent work	13. Share of women in workforce	%
		14. Share of women in management positions	%
		15. Wages at lowest wage group	$
Economic	(f) Financials	16. Net profit margin	%
		17. Return of capital employed	%
	(g) Development	18. Investment in R&D	$
		19. Investment in staff development	$
		20. Expenditures on EHS compliance	$

Table 3.3 Judging of key performance indicators

Indicators of positive sustainability		Indicators of negative sustainability	
Safety trainings	Net profit margin	Energy use	Working accidents
Training and education	Investment in R&D	Material use	Sickness frequency
Share of women in workforce	Investment in staff development	Freshwater consumption	Employee attrition rate
Share of management positions		Waste generation	Hazardous materials
Wages at lowest wage group		GWP	Expenditures on EHS compliance
Return of capital employed		AP	

3.3.3 *Judging of Sustainability Performance Indicators*

After identifying the key indicators, it is necessary to determine whether an indicator has a positive or negative influence on a company's sustainability performance. This judgment of each indicator becomes important for the normalization and the aggregation in the next steps. Positive indicators are considered as sustainability contributing and should therefore be maximized. Negative indicators on the other hand should be minimized to support sustainability. Thus, the results are summarized in Table 3.3.

3.3.4 *Normalizing of Sustainability Performance Indicators*

The next step toward calculating a composite sustainability performance index focuses on the normalization of indicators. This step is important because the indicators are expressed in different units, and the combination of the indicators into the performance index requires common units to achieve a representative result. A number of normalization methods exist in the literature, and the main procedures are presented in the following.

3.3.4.1 Normalization Methods

Minimum–Maximum

This method normalizes indicators with a positive impact on sustainability by the equation:

$$I^{+}_{N_{ijt}} = \frac{I^{+}_{ijt} - I^{+,Min}_{ij}}{I^{+,Max}_{ij} - I^{+,Min}_{ij}} \quad (3.1)$$

Indicators with a negative impact on the other hand are normalized by the equation:

$$I^{-}_{N_{ijt}} = \frac{I^{-}_{ijt} - I^{-,Min}_{ij}}{I^{-,Max}_{ij} - I^{-,Min}_{ij}} \quad (3.2)$$

where $I^{+}_{i,j,t}$ and $I^{-}_{i,j,t}$ are the values for indicator i from the group of indicator j in year t with positive and negative impacts on sustainability, respectively, while $I^{+}_{N,i,j,t}$ and $I^{-}_{N,i,j,t}$ are the normalized positive and negative indicators, respectively. Overall, this transformation results in a clear compatibility of different indicators, but it requires a valid database in order to be carried out (Zhou et al. 2012; OECD 2008).

Distance to a reference
This method calculates the ratio between the indicator and an external benchmark. The normalized indicators are described in the following equation:

$$I_{N_{ij}} = \frac{I_{ij}}{I^{Benchmark}_{ij}} \quad (3.3)$$

$I^{Benchmark}_{ij}$ is the benchmark for indicator i from the group of indicators j. In this case, it is possible that the normalized value is higher than 1, which indicates that the performance of the factory is better than benchmark. Similar to the previous method, this approach can only be carried out if more than one set of data is available (Zhou et al. 2012; OECD 2008).

Percentage over annual differences
Finally, the method "percentage over annual differences" is the third main normalization approach discussed in this paper. It focuses on the development of the indicators over time. Therefore, each indicator is transformed using the following formula:

$$I_{N_{ijt}} = \frac{I_{ijt} - I_{ij,t-1}}{I_{ij,t-1}} * 100 \quad (3.4)$$

Nevertheless, the disadvantage of this method concerns the case $t = t^{o}$. In that case, the indicators cannot be normalized by the given equation and the data would be lost during the analysis (Zhou et al. 2012).

3.3.4.2 Evaluation of Normalizing Method for Factory Sustainability

All of the described methods require a database or a set of reference data in order to transform the indicators. However, since there is no database available for the indicators, normalization is not possible for one data set of indicators. Nevertheless, this tool does attempt not only to assess a single factory but also to compare different factories with each other and to evaluate the development over time. These three different cases lead to the following conclusion:

- Assessing a single factory → no normalization possible.
- Comparing different factories → "distance to a reference,"
- Development of a factory → "percentage over annual differences."

3.3.5 Weighting of Sustainability Performance Indicators

The next step focuses on weighting the indicators in order to determine the individual importance of the indicators toward an overall goal. Although this purpose is understood easily, it remains difficult to determine weights with sufficient accuracy (Krajnc and Glavič 2005). "The relative importance of the indicators is a source of contention" (OECD 2008, p. 33). Therefore, a number of different weighting techniques exist in the literature. Some are derived from statistical methods such as the data envelopment analysis and others from participatory methods like the analytic hierarchy process.

3.3.5.1 Weighting Methods

Budget Allocation Process (BAP)
This weighting procedure determines the indicator weights based on expert opinion. In general, the BAP has four different phases. First, experts in the field have to be selected for the assessment. It is essential that the experts represent a wide spectrum of knowledge and experience. Second, the selected experts have to allocate a "budget" of one hundred points to the indicator set, based on their personal judgment of the relative importance. In a third step, weights are calculated as average budgets. As an optional fourth step, the procedure could be iterated until convergence is reached (Hermans et al. 2008; OECD 2008).

The main advantages of BAP are its transparent and simple application as well as its short duration. On the other hand, it also contains several disadvantages: The weights are fairly subjective and could reflect specific conditions that are not transferable from one factory to another (Zhou et al. 2012).

Analytic Hierarchy Process (AHP)
The analytic hierarchy process is another participatory method similar to the budget allocation process. However, this method is far more complex and consists of a mathematical approach. As a first step, it is necessary to translate a complex problem into a hierarchy. The second step requires a pair-wise comparison between each pair of indicators. Experts have to judge "how important is indicator j relative to indicator i ?". Values on a scale from 1 to 9 are assigned to show the intensity of preference

(Saaty 1980). In the next step, the results are presented in a matrix to obtain the relative weights of each indicator (OECD 2008). Finally, it is required to find the eigenvector with the largest eigenvalue from the matrix. The eigenvector presents the weights, and the eigenvalue evaluates whether the judgment is consistent or not (Saaty 1980; Singh et al. 2007).

Aside from the problem of possible inconsistency, the subjectivity of judgment is another negative characteristic of the method. Each expert judges the indicators based on his or her own knowledge and experiences. Thus, the possible inconsistency is also related to subjectivity (Hermans et al. 2008). Despite these disadvantages, AHP is a comprehensive and popular technique, and the information from well-selected experts is valuable for weighting indicators. Singh et al. used this method to develop a composite sustainability performance index for the steel industry (Singh et al. 2007), Krajnc et al. applied it to a case study on the sustainability performance of the oil and gas industry (Krajnc and Glavič 2005), and Hermans et al. implemented it to a limited extent in the road safety research (Hermans et al. 2008). In contrast to most other methods, AHP allows both, quantitative and qualitative criteria, to be entered into the model and it assesses different levels of criteria.

Data Envelopment Analysis (DEA)
The data envelopment analysis, developed by Charnes, Cooper, and Rhodes (CCR) in 1978, is a linear programming method that can be used for calculating the relative efficiency of decision-making units (DMUs). In the context of this study, each factory can be considered as a DMU.

The DEA is different compared to the other weighting methods. The model results in factory-specific weights instead of one set of weights for all factories. This is a disadvantage because factories can only be ranked and compared if they are based on the same set of weights. Furthermore, in this approach the weights do not sum up to 1, which makes the comparison with other weighting methods difficult. Nevertheless, this method has already been used for a number of indices such as the technology achievement index (Cherchye et al. 2007). The strength of the method lies in the fact that the optimal weights are directly derived from the data and that no normalization is needed.

Benefit-of-the-doubt (BOD)
The benefit-of-the-doubt presents another application of the DEA in the field of composite indicators. In contrast to the original DEA model, BOD evaluates the relative performance of the factories and not the efficiency (Cherchye et al. 2004). However, it is based on the same model and follows the same process. The composite index CI in this case is calculated as the ratio between the actual performance of the factory and the external benchmark.

Since BOD can be seen as a specialized version of the original DEA model, the DEA's advantages and disadvantages also apply to this method. However, this method has already been used for a number of indices. It was originally proposed in the context of a macroeconomic performance assessment by Melyn and Moesen in 1991 and later adapted by Cherye and Kuosmann for a cross-country assessment of human development and sustainable development performance (Cherchye et al. 2004).

Equal Weighting (EW)

As its name already indicates, the same weight is assigned to each indicator. This implies that all indicators have the same importance and that no statistical or participatory approach is used to determine the weights. The value of the weights is simply calculated by $\frac{1}{l}$ where l is the number of all indicators and 1 represents the sum of all weights (Zhou et al. 2012; Hermans et al. 2008).

Although this method appears too simple from a scientific point of view, several composite indicators like the environmental sustainability index or the European Innovation Scoreboard are constructed by equal weighting (Hermans et al. 2008). The main disadvantage is the fact that it does not offer any insights into indicator importance and it does not reflect reality. However, this method can be considered as a solution in case no other weighting method presents valid results.

3.3.5.2 Selection of a Weighting Method for Factory Sustainability

In order to analyze which weighting method is best fitted and suitable for a framework to assess sustainability of manufacturing companies, it is required to develop specific criteria that have to be fulfilled.

- *Quantitative and qualitative data*: Since the set of indicators that are used for this framework may be extended by qualitative indicators, it is necessary that the weighting method can handle both types of data.
- *Objectivity*: Indicators should be weighted without bias in order to be meaningful and to decrease personal preferences.
- *Insights into indicator importance*: The overall goal of the tool developed within this study is to assess the sustainability performance of manufacturing companies. Therefore, it is important that indicators reflect their individual importance toward company sustainability.
- *Transferability*: The developed tool is supposed to allow the user to compare companies with each other. In order to do so, it is required that the indicator-specific weights are always valid and transferable from company to company.
- *No need for a database*: Due to the fact that there is no large accessible database for each indicator, the weighting has to be possible without including a lot of data.

In order to identify a weighting method, the methods introduced previously are presented in a structured way in Table 3.4, where each method is assessed toward the fulfillment of the before-derived criteria.

A method fulfilling all the criteria cannot be identified. However, of all the reviewed weighting methods the AHP gives the best results. It is therefore used to weight the different indicators within this study.

Table 3.4 Evaluation of weighting methods

Method	Quantitative/ qualitative data	Objectivity	Insights into indicator importance	Transferability	No need for a database
BAP	+	–	O	O	+
AHP	+	O	+	+	+
DEA	O	+	O	–	–
BOD	+	+	O	–	–
EW	+	–	–	+	+

3.3.6 *Calculating the Subindices and Composite Index*

After weighting and normalizing each indicator, the next step requires grouping them into subindices for each group of sustainability indicators. In the context of this study, there are three groups of indicators (environmental, economic, and social) and therefore also three subindices, respectively. They can be derived as shown in the following equation:

$$I_{S_j} = \sum_{ji}^{n} w_{ji} * I_{N_{ji}} \quad (3.5)$$

$$\text{s.t.} \sum_{\text{ji}}^{\text{n}} \text{w}_{\text{ji}} = 1, \text{w}_{\text{ji}} \geq 0 \quad (3.6)$$

where SI_j is the sustainability subindex for each group of indicators j. Since the framework uses the AHP weighting method, the first constraint restricts the sum of all weights w_ji of indicator i for the group of sustainability indicators j to be equal to 1.

As a final step, it is required to combine all three subindices into one overall composite sustainable performance index.

$$CI = \sum_{j=1}^{n} w_j * I_{S_j} \quad (3.7)$$

$$s.t. \sum_{j=1}^{n} w_j = 1, w_j \geq 0 \quad (3.8)$$

where CI is the overall sustainability composite index for the factory that has been assessed.

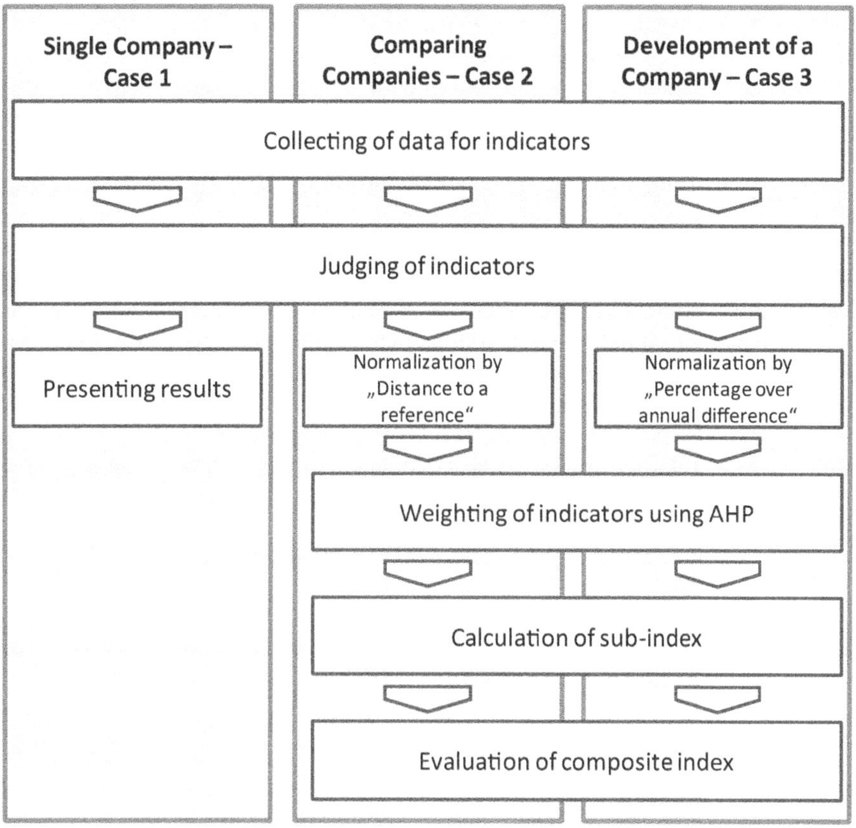

Fig. 3.4 Scheme of the framework

3.3.7 Framework of the Factory Sustainability Assessment Tool

Regarding the developed framework, it shows that three different cases can be evaluated using this assessment tool (see Fig. 3.4).

First, it is possible to consider only a single manufacturing company. The collected data for this case can be judged according to the results of Sect. 3.3.4, but since there is no reference data available, normalization cannot be performed. This process ends here with the presentation of the results. The next case regards the comparison of two or more manufacturing companies. After collecting the relevant data for each indicator, they can be normalized by using the method "distance to a reference." However, some aspects of this method have to be slightly modified in order to meet the requirements within this study. As there is no large database available, there is also no external benchmark value available. Although it is not possible to normalize one company with a benchmark value, it is still possible to compare two or more companies by assigning the value 1 to the inferior company for each indicator and therefore making it a reference company. The remaining manufacturing companies are evaluated relatively to that company with a value between 0 and 1. The closer the value is to 0, the better the performance of the company according to that indicator. In the next step, each indicator can be weighted according to its importance toward the overall goal by using the "analytic hierarchy process." Using this method, a group of sustainability experts with different industrial backgrounds were asked to judge the indicators by estimating a preference factor of each indicator relative to another. To ensure the consistency of the judgment, a consistency ratio was calculated and confirmed the assumption. Afterward, the normalized and weighted indicators can be combined to a subindex and then to an overall composite index. The third case considers the development of a company over time. This case is similar to the second case. However, this case uses the normalization method "percentage over annual differences." It is the best-suited method because only one company is considered over time. Nevertheless, if only a few data sets are available, it is recommended to use the same normalization method as explained above, because the data for $t = t^o$ would be lost during the analysis.

3.4 Case Study

After implementing the framework into a computer-based tool, it has been applied to a practical case study in order to demonstrate its usability and effectiveness. The case study is divided into two parts. The first part focuses on the comparison of different manufacturing companies (Case 2) by analyzing three leading automotive manufacturers. The second part of this case study considers the assessment of a manufacturing company over time (Case 3). In this case, the results of one automotive manufacturer are compared for the years 2010, 2011, and 2012. However, the assessment of a single company (Case 1) is not considered in this chapter because it is already included, while Case 2 or Case 3 is performed. The entire case study is carried out by using the developed sustainability assessment tool and data based on public available records.

3.4.1 Comparison of Different Manufacturing Companies

This section of the case study evaluates the sustainability performance of three different companies from the automotive industry that produce a similar product in the period from January 1 to December 31, 2013. For the most part, the data were taken from sustainability reports, because all data in these reports were audited and verified by a third party and are therefore a reliable source. Other data had to be taken from press releases or Web pages. The data entry and data processing are then carried out by using the developed assessment tools.

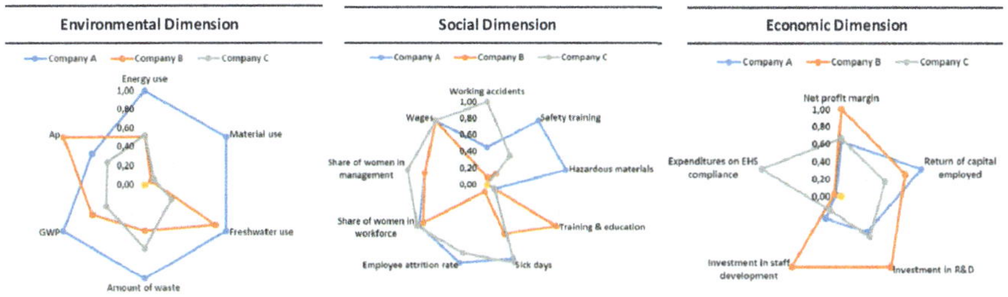

Fig. 3.5 Presentation of assessment results by environmental, social, and economic indicators—Case 2

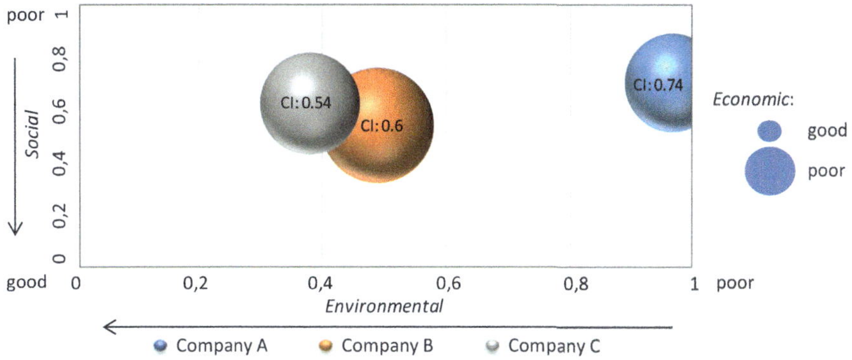

Fig. 3.6 Presentation of assessment results by dimensions—Case 2

In the first step of the analysis, the results of each indicator are illustrated in a spider chart and are compared for all three companies. The results of this comparison are presented in the following figure. The closer the value is to "0" the better the performance. The results of the comparison are presented in Fig. 3.5.

This comparison demonstrates that company C performs better than company A and B in terms of most environmental indicators. Also, it can be determined that company A shows the weakest performance, except for the *acidification potential*. Regarding the social and economic dimension, the results are not that as clear. The performance varies from indicator to indicator. However, it can be seen that company B has a general trend to perform better regarding the social indicators and weaker regarding the economic indicators. In order to make a general statement about the sustainability performance, the results of each indicator have to be weighted and combined into a composite index. The results of this process are visualized in Fig. 3.6.

Regarding the subindices, it can be concluded that company C performs the best and company A the weakest in terms of the environmental dimension. Regarding the social dimension, company B performs the best, but the worst regarding the economic dimension. However, looking at all three dimensions the overall composite index (CI) displays that company C shows the best results. This clear and comprehensible

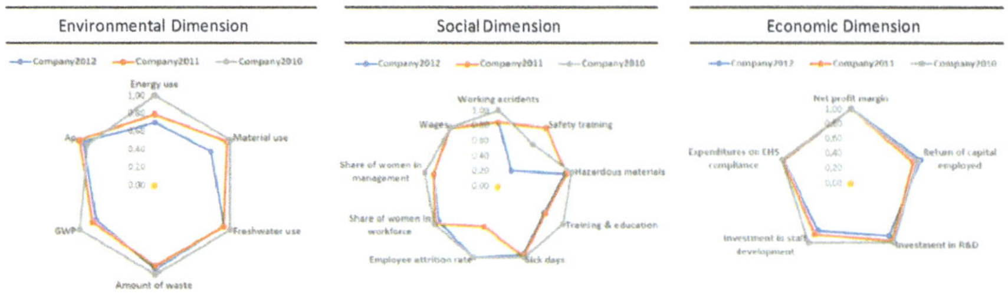

Fig. 3.7 Presentation of assessment results by environmental, social, and economic indicators—Case 3

presentation of results might be interesting for a factory manager of either one of the three companies in order to improve the performance in the future. On the other hand, these results are also interesting for any investor with a focus on sustainable investing.

3.4.2 Assessment of a Manufacturing Company over Time

In this section, the sustainability performance of only one company from the automotive industry is assessed over time. For this case study, the evaluation period for all three periods is again the calendar year from January 1 to December 31. Again, all data and information are gathered from public records, e.g., environmental declarations, Web pages, and sustainability reports.

After gathering and entering the data for each evaluation period, the data processing is performed. According to the previous section, the first step of the data processing considers the data for each indicator and compares them for each evaluation period. The results are shown in Fig. 3.7.

The spider charts present clearly the development of each indicator. With regard to the environmental dimension, it can be seen that the sustainability performance for almost every indicator improves steadily over time. Only the indicators *acidification potential* and *waste generation* show minor differences. Regarding the social dimension, it is significant that the *employee attrition rate* has increased from 2011 to 2012. Furthermore, it is striking that *safety trainings* show a discontinuous development. Additionally, all indicators concerning the economic dimension are rather unremarkable and present only minor differences.

In the next step of this case study, it is important to analyze the weighted and combined subindices. The results are visualized in Fig. 3.8. Recalling that the closer the index is to 0 the better the sustainability performance, it can be determined that the environmental and social indices have improved significantly over time. However, the economic dimension shows a minor difference. Here, the index for the year 2011 is slightly better than the index for the year 2012. Nevertheless, the overall composite

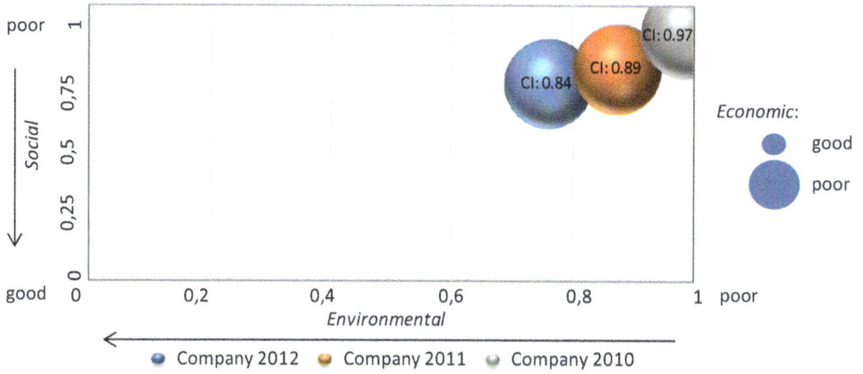

Fig. 3.8 Presentation of assessment results by dimensions—Case 3

sustainability index indicates again the continuous improvement of the sustainability performance from CI 0.97 to CI 0.84 over the last 3 years.

As a conclusion of this case study, it can be summarized that the general development of the sustainability performance demonstrates a continuous improvement. However, it shows also that in the future the factory managers should focus more on economic indicators and also on the employee attrition rate.

3.4.3 Results of the Case Study

The case study has been carried out without any significant complications. The results of the study are clearly visualized and provide detailed information that might be used by factory managers as well as investors to support decisions and to guide future activities.

However, the data collection based on public records required more time than expected. Generally, sustainability reports provide most of the data for each environmental indicator and also for some social and economic indicators. Other indicators had to be sought in press releases or Web pages, and especially the data collection for the indicator "wages" has proven to be difficult. In this case, it is, for example, necessary to refer to wage agreements of trade unions. Nevertheless, if detailed sustainability reports are not available for smaller companies, the time effort to collect all data increases enormously. Besides the problem of data availability, most companies publish their figures and data with different units. Therefore, it is sometimes necessary to convert the units, whereby accuracy might be lost. However, after collecting all numbers and figures, data entering and processing can be performed with minimal time and effort.

Overall, it can be concluded that data collection for large companies with a focus on sustainability is significantly easier than for small and individual production sites.

However, since sustainability is attracting more and more attention, it becomes also more important for companies to be certified by environmental audits such as Eco-Management and Audit Scheme (EMAS) and therefore they have to publish more figures and data in the future. For now, it cannot be guaranteed to find all data on public records, but there is always the option to get in touch with the sustainability contact person in order to gather more information about a specific production site.

3.5 Conclusion and Outlook

In this study, the focus is on developing a new integrated and rapid sustainability assessment tool at the level of manufacturing companies and from an external and internal perspective. Therefore, a categorization of existing assessment tools as well as the requirements for a new tool is described. To meet the requirements, the paper presents a method to develop an integrated framework which is especially designed to assess a manufacturing company and considers all three dimensions of sustainability. Using this framework, it is possible to assess a single company, to compare different companies, and to assess a company over time. A practical case study confirms that the tool needs less than 30 min for data entering and data processing and it can be mostly completed without internal information. It can be concluded that the tool provides clear results and presents them in a comprehensible form. Based on these results, the tool supports investor's decision on sustainable investing and it may also guide factory managers to think and act in the right direction and to discover possible improvements in order to increase the sustainability metrics related to factory operations.

However, there is still potential for future research on this topic. The significance of composite indicators should be further discussed, as they may provide misleading and non-robust messages. A single indicator is heavily summarized so that it does not allow any detailed conclusions on the results. Therefore, a sensitivity analysis of the composite index might be useful and the subindex as well as the single indicators should be considered as well. Further, the framework of the tool provided in this work does not offer the calculation of an index for the assessment of a single company. This is due to the fact that no database or standardized scale for the assessment values is available and the values cannot be normalized. This issue might be solved by collecting data for the leading company in each industrial sector. In case a single company is being assessed, the results can be normalized relative to the benchmark company of the specific industry. Another step of further improvement considers the weighting of indicators by using the AHP method. Since it is a participatory method, the results will be more sophisticated the more experts participate. In the future, it might also be possible to place the entire tool on a public Web page. Therefore, different users have access to it and they would be able to share a database. The variety of factories in the database would increase, and the users have the opportunity to compare the results of different factories with a minimal amount of effort. This may also solve the problem concerning the complicated and time-consuming data collection for small and midsize factories based on public records.

Acknowledgements This research is based on a master thesis jointly supervised by the University of Rhode Island and the Technische Universität Braunschweig. The full thesis is available here: http://digitalcommons.uri.edu/cgi/viewcontent.cgi?article=1135&context=theses.

References

Chen D, Schudeleit T, Posselt G, Thiede S (2013) A state-of-the-art review and evaluation of tools for factory sustainability assessment. Procedia CIRP 85–90

Cherchye L, Moesen W, Rogge N, van Puyenbroeck T, Saisana M, Saltelli A, Liska R, Tarantola S (2007) Creating composite indicators with DEA and robustness analysis: the case of the Technology Achievement Index. J Oper Res Soc 59:239–251

Cherchye L, Moesen W, van Puyenbroeck T (2004) Legitimately diverse, yet comparable: on synthesizing social inclusion performance in the EU. JCMS: J Common Mark Stud 42:919–955

Davidson CI, Hendrickson CT, Matthews HS, Bridges MW, Allen DT, Murphy CF, Allenby BR, Crittenden JC, Austin S (2010) Preparing future engineers for challenges of the 21st century: sustainable engineering. In: Going beyond the rhetoric: system-wide changes in universities for sustainable societies, vol 18, pp 698–701

DJSI (2013) Dow Jones Sustainability Index. http://www.sustainabilityindex.com/dow-jones-sustainability-indexes/index.jsp/. Accessed 13 Apr 2013

Feng SC, Joung C-B (2009) An overview of a proposed measurement infrastructure for sustainable manufacturing

Feng SC, Joung C-B, Li G (2010a) Development overview of sustainable manufacturing metrics. http://www.sustainelectronics.illinois.edu/NSFworkshop/Reading/Development%20Overview%20of%20Sustainable%20Manufacturing%20Metrics.pdf. Accessed 2 June 2013

Feng SC, Joung C-B, Li G (2010b) Development overview of sustainable manufacturing metrics. In: Proceedings of the 17th CIRP international conference on life cycle engineering, Hefei, China

Global Reporting Initiative (2011) Sustainability reporting guidelines, Netherlends. https://www.globalreporting.org/resourcelibrary/G3.1-Guidelines-Incl-Technical-Protocol.pdf. Accessed 17 Apr 2013

Graedel TE, Allenby BR (2010) Industrial ecology and sustainable engineering, International ed. Pearson, Boston

Gutowski TG, Allwood JM, Herrmann C, Sahni S (2013) A global assessment of manufacturing: economic development, energy use, carbon emissions, and the potential for energy efficiency and materials recycling. Annu Rev Environ Resour 38:81–106

Hermans E, van den Bossche F, Wets G (2008) Combining road safety information in a performance index. Accid Anal Prev 40:1337–1344

Krajnc D, Glavič P (2005) How to compare companies on relevant dimensions of sustainability. Ecol Econ 55:551–563

Labuschagne C, Brent AC, van Erck RP (2005) Assessing the sustainability performances of industries. J Clean Prod 13:373–385

Ness B, Urbel-Piirsalu E, Anderberg S, Olsson L (2007) Categorising tools for sustainability assessment. Ecol Econ 60:498–508

OECD (2008) Handbook on constructing composite indicators. Methodology and user guide. OECD Publications, Paris

Ranganathan J (1988) Sustainability rulers: measuring corporate environmental and social performance. http://pdf.wri.org/sustainability_rulers.pdf. Accessed 2 June 2013

Saaty TL (1980) The analytic hierarchy process. Planning, priority setting, resource allocation. McGraw-Hill, New York
Singh RK, Murty HR, Gupta SK, Dikshit AK (2007) Development of composite sustainability performance index for steel industry. Ecol Ind 7:565–588
Srinivasan V (2011) An Engineer grapples with sustainable manufacturing. Clean Technol Environ Policy 13:217–219
Sustainable Measures (2010) Characteristics of effective indicators: sustainable measures. http://www.sustainablemeasures.com/node/92. Accessed 1 Sept 2014
Veleva V, Ellenbecker M (2001) Indicators of sustainable production: framework and methodology. J Clean Prod 9:519–549
WCED—World Commission on Environment and Development (1987) Our common future (The Brundtland-Report). Oxford
Zhou L, Tokos H, Krajnc D, Yang Y (2012) Sustainability performance evaluation in industry by composite sustainability index. Clean Technol Environ Policy 14:789–803

Chapter 4
Sustainability Assessment in Manufacturing and Target Setting in Highly Automated Production

Jan-Markus Rödger and Niki Bey

Abstract The interest in sustainability in manufacturing is growing. The European Commission elaborated in collaboration with the European Factories of the Future Research Association (EFFRA) a road map until 2030. The revised ISO 14001 standard, released end of 2015, emphasizes the integration of a life cycle perspective in production. External stakeholders like the Carbon Disclosure Project are striving for very detailed sustainability-related information. Therefore, this chapter will describe the latest approaches of how to assess sustainability in manufacturing, how production planning works in automated production and how sustainability thinking might be integrated in a feasible way. Additionally, a new framework is described which is able to link and allocate the product-related life cycle emissions in a consistent way from the large context down to the individual machine tool level in manufacturing. In essence, this chapter points out that efforts in manufacturing improvements need to be done with a view on entire systems rather than with a view only on single "island solutions"—and it shows a way how to do this.

4.1 Introduction

In Europe, the share of manufacturing of the gross domestic product (GDP) was 15% (equals 17.7 trillion USD) in 2011, employing 30 million people (about 16% of the working population) according to the ISIC Rev. 4 statistics of the OECD and World Bank (2011a, b). Irrespective of this relatively small share of societal activity, the environmental impact of manufacturing activities needs to decrease substantially in the future, for two reasons:

Firstly, when applying generic sustainability equations such as $I = P * A * T$ (Graedel and Allenby 1995), which related overall "Impact" (I) to the product of

J.-M. Rödger (✉) · N. Bey
Division of Quantitative Sustainability Assessment, Department of Management Engineering, Technical University of Denmark (DTU), Produktionstorvet, Building 424, 2800 Kongens Lyngby, Denmark
e-mail: januw@dtu.dk

S. Thiede and C. Herrmann (eds.), *Eco-Factories of the Future*, Sustainable Production, Life Cycle Engineering and Management,
https://doi.org/10.1007/978-3-319-93730-4_4

"Population growth" (P), "Affluence" (A), i.e. "Standard of living", and "Technology" (T), the two factors P and A will increase steadily in value. This leaves the factor "Technology" (T)—and hereunder "manufacturing" in general including "highly automated production" in particular—as the only "lever" left to generate the required substantial reductions in overall Impact (I). This reduction of overall impact needs to be done in a balanced approach considering environmental protection, economic growth and social well-being—also referred as "the three pillars of sustainability"—in order to be in accordance with the generally accepted definition of sustainable development coined in 1987 by the so-called Brundtland Commission. Secondly, when extrapolating current developments in deployment of renewable energy sources and resulting reductions in product-use-stage impacts, it can be expected that impacts of manufacturing will become the dominating ones in the life cycles of any electricity-intensive product type. For example, in 2010 the manufacturing sector contributed with around 19% to the worldwide carbon dioxide (CO_2) emissions (World Bank 2011a, b). Between 2005 and 2010, these CO_2 emissions increased on average by 2.8% per year. In the same period, the population size increased by 1.2%, and the GDP per capita (i.e. the factor "Affluence") increased by 5.6% (inflation-adjusted) annually.

In recognizing the described special influence that manufacturing can have on the reduction of overall impact and thus on improving sustainability performance, this chapter argues for the key understanding that efforts in manufacturing improvements need to be done with a view on entire systems rather than with a view only on single "island issues". Failing to look at entire systems may provoke sub-optimizations or—in the worst case—even lead to directly unintended outcomes, such as problem shifting, e.g. from one impact category to another one (e.g. less CO_2 but more chemicals) or from one life cycle stage to another one (lower impacts during use but higher impacts in end-of-life). Possible improvements, only visible from the suggested "helicopter perspective" that covers the entire system, may otherwise not be identified and utilized—and the actual problem may not only remain unsolved but may unintentionally even be made worse.

In addition, assessing highly automated production systems within such a holistic context also yields potential for reaching win-win situations. Assessment results can be communicated as response to increasing company-external pressures, such as market and legislative requirements, and at the same time may release earlier not identified improvement potentials, which is a characteristic company-internal interest.

In addressing the challenge of making such holistic assessments, this chapter first gives a brief overview of main approaches suggested of sustainability performance assessment, then discusses sustainability aspects particular to automated production and further describes the way in which sustainability issues currently are included in typical decision-making processes for automated car production. Based on this, a key section introduces a new framework for sustainability assessment of highly automated productions, and the chapter closes with a number of conclusions and suggestions for further work.

4.2 Approaches Towards Holistic Sustainability Assessment

In the context of this chapter, the term "sustainability assessment" covers any systematic way of analysing and assessing human-activity-caused potential adverse effects on the environmental, social and economic sustainability of a system. Such a system would typically be a product system, covering all life cycle stages of that product (thus also any manufacturing activities) including all ingoing and outgoing flows crossing the system's delimitations.

The theoretical foundations for sustainability assessment were laid in the 1960s by, for the first time, establishing cause–effect chains, i.e. relationships between emissions to the environment (causes) and effects in the environment. Focus was on single environmental causes, such as certain chemical pesticides and their harmful (side) effects in the environment—a prominent example being the insecticide DDT and the identification of its potential for human-health-impairing effects described by Rachel Carson in her book "Silent Spring" (Carson 1962). Already this example demonstrated the inherent issue that sustainability assessments typically have to cope with trade-offs consisting in the fact that the desired effects of the particular human activity may have unwanted side effects. In the DDT case, the intended protection of farming crops from destruction by insects had the unintended side effect of human-health impairments of people being exposed to the insecticide. Since such trade-offs are characteristic for sustainability assessments, they need to be identified and addressed by any practitioner.

Building on this, multi-issue approaches emerged such as Environmental Impact Assessment (EIA), Environmental Risk Assessment (ERA) and Life Cycle Assessment (LCA) (Hauschild 2005). LCA is ISO-standardized (in ISO 14040/44) and is the dedicated approach to assessing diffuse, i.e. not location-specific, sources through introducing the product life cycle perspective. Instead of "just" covering single sources (e.g. effluents from a production site), the life cycle perspective allows assessing diffuse multiple sources—this is enabled by looking at and in the assessment always relating to the "cause" of emissions, which is the product (or service) and the activities (processes) taking place over the product life. Taking such a life cycle view is referred to as "Life Cycle Thinking", and the life cycle approach is the backbone in relevant legislation and standardization.

The "life cycle" of a product covers all stages during the products' useful life; starting with extraction of raw materials (e.g. iron ore) and production of materials (e.g. steel), over manufacturing of the product (e.g. a car manufactured in a factory), use and maintenance of the product (e.g. driving a car) to end-of-life (e.g. landfilling, incinerating or recycling of the car). An advantage of using a life cycle approach is that shifting problems from one part of the life cycle to another part will not go unnoticed. Burden-shifting can thus be avoided, including potential unintended shifts from one impact type, e.g. energy-related emissions, to another type (e.g. toxic substances) since the view is always on the entire product life cycle system. In order to make comparisons of different products possible, incl. their individual entire life cycle systems, any LCA requires the definition of a so-called Functional Unit, a description of the functionality that the product (or service) fulfils. Any comparison has to be done on the same Functional Unit, i.e. the functionality fulfilled by the compared product life cycle systems has to be the same.

Among the life cycle-based approaches, the methodology developed furthest is Life Cycle Assessment, LCA, allowing assessment of the environmental sustainability of a product system within defined system boundaries. Applying the same system boundaries, Life cycle costing (LCC, e.g. UNEP/SETAC 2011) can be used to map all costs and revenues in the life cycle of a product. Social or Societal LCA (SLCA, e.g. Jørgensen et al. 2010) covers the third pillar of sustainability. The latter method is still very young and under development, but it aims to cover the positive and negative social and socio-economic aspects along the life cycle of a product (Du et al. 2015).

While most approaches towards sustainability assessment in fact focus highly or even exclusively on assessing the environmental dimension of sustainability, developments are ongoing towards *holistic* assessments, i.e. towards life cycle system-wide assessments that consider all three sustainability pillars. An example is "Life Cycle *Sustainability* Assessment" (LCSA, e.g. Finkbeiner et al. 2010) which combines the methods LCA, LCC and SLCA.

The above methodological approaches inevitably produce complex, multidimensional results. An effort to operationalize result presentation was made by Traverso et al. (2012): in their so-called sustainability dashboard, ten indicators in each of the pillars of sustainability were selected. The indicators chosen by the developers of the sustainability dashboard are shown in Table 4.1. In order to arrive at an overall, single-figure sustainability score, a method to weigh each indicator, was suggested.

In another principle effort of improving the communication of such multidimensional results and support decision-making regarding individual key issues, simplified "footprinting" approaches are established. Most of them keep the product life cycle perspective, while focusing on single impact categories (e.g. global warming). Examples are Carbon Footprint (BSI 2011), Product Environmental Footprint (EU JRC 2014) and Water Footprints (Hoekstra 2010).

A prominent emerging type of sustainability assessment frameworks particularly recognizes the finite boundaries of Planet Earth (Planetary Boundaries, Absolute Sustainability). As opposed to assessing product systems and measuring relative improvements (under a regime of eco-efficiency), these emerging frameworks suggest assessing product systems (and services/activities) against absolute boundaries, thus particularly including "rebound effects" and other important effects, which otherwise are not considered. A "rebound effect" is the overall adverse effect occurring, when a product generation is improved relative to a previous product generation (e.g. by 10% in a given impact category) and when the producer then sells for instance twice as many units of this improved generation—meaning that the overall impact caused by the new product generation still increases. Examples for this type of approach are "Ecological Footprint" (Borucke et al. 2013), "Blue Water Footprint" (Hoekstra and Mekonnen 2012), "Planetary Boundaries" (Rockström et al. 2009; Steffen et al. 2015) and "Absolute Sustainability" (Bjørn and Hauschild 2013).

Table 4.1 Indicators used in the "Sustainability Dashboard" (Traverso et al. 2012)

Sustainability pillar	Background method	Indicators
Environment	Life Cycle Assessment, LCA	Embodied energy, global warming, human toxicity, photochemical oxidation, acidification, eutrophication, abiotic depletion, ozone layer depletion, terrestrial ecotoxicity
Economy	Life cycle costing, LCC	Extraction costs, manufacturing costs, finishing costs, waste disposal costs, electricity costs, equipment costs, revenues, fuel costs, raw material costs
Society	Social LCA, SLCA	Salary per employee, percentage of female workers, percentage of females at the administration level, percentage of employees with limited contracts, percentage of workers with yearly check up, number of accidents, percentage of child labour, number of discrimination cases, social benefits per employee

Related to manufacturing, there are numerous additional approaches, indicator sets and principles available to support companies in assessing their sustainability performance (Amrina and Yusof 2011; Schrettle et al. 2014; Warhurst 2002). Duflou et al. (2012) described a structured approach for energy-related aspects where different system levels were distinguished: from a unit process over a multi-machine, a factory, a multi-facility to supply chain level. For instance, Garetti and Taisch (2012) have considered education and technology as additional factors in sustainable manufacturing.

The European Commission elaborated in collaboration with the European Factories of the Future Research Association (EFFRA) a road map in 2013 of factories in 2030, addressing several key aspects like lower resource consumption and sustainability in production processes to improve the competitiveness (EFFRA 2013). According to Thiede (2012), sustainable manufacturing might lead to some drawbacks for the manufacturability. Baldwin et al. (2005) and Bey et al. (2013) address that not the lack of strategies, models and tools are the main barriers, but how to implement them and more importantly how to introduce them into existing practices. To solve these concerns, an understanding of relevant aspects in production is needed and will be described in the following paragraphs.

4.3 Sustainability Aspects of Automated Production

The integration of economic, environmental and social requirements, the development of innovative products and services, and the comprehensive usage of available knowledge are core factors for business success and sustainable development (Herrmann 2010). The car industry, as one of the most automated sectors in particular (IFR 2013), which acts mostly on the international market, always struggles to predict the actual market demand (the so-called functionality) and is affected by several international and national regulations. This implies new chances but also new risks, uncertainties and dynamic changes. Each market has its own requirements towards the life cycle of a product, which can be expressed in quantitative way using the concept of a "Functional Unit" (FU) of a product, which is part of LCA methodology (see Sect. 4.2 of this chapter).

Today, total turnover and revenue are the key indicators for the companies and their stakeholders. But the market already asks for more sustainable products, which is reflected by, e.g., new CO_2 emission regulations (e.g. EU 2014). Although it is not statutory, companies assessing their products over the entire life cycle by using Life Cycle Assessment standards like ISO 14040/44 (2006). In those LCAs, a car is analysed over its entire life cycle in detail to assess different potential environmental impacts (e.g. global warming, eutrophication) during the manufacturing, use and end-of-life (EoL) stage.

The main car manufacturers publish simplified LCAs of their latest products to inform the customers or use it for internal purposes. Based on 21 publicly available analyses, it can be stated that around 17% of the carbon dioxide emissions are due to manufacturing, the use-stage accounts for 82% and the EoL stage only 1% to the overall life cycle (see Fig. 4.1). If all other greenhouse gases (e.g. methane) would have been reported consistently in the LCAs, the share of manufacturing would even increase slightly.

Improvements were achieved in lowering the specific fuel consumption of road vehicles over the last two decades (OECD/IEA 2012). This trend will be most likely amplified by additional research and development efforts for new technology like

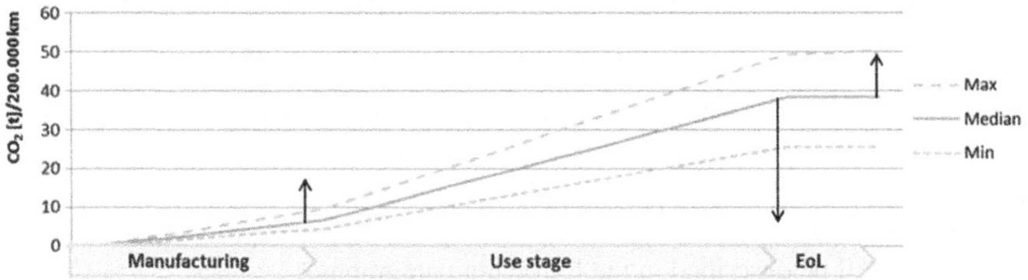

Fig. 4.1 Comparison of cumulative carbon dioxide emissions (CO_2) during manufacturing, use and end-of-life based on 21 different LCAs published by car manufacturers and normalized to 200.000 km travel distance and prospective development of the stages (black arrows)

electrified or hydrogen cars, with a prospective share of around 10% in 2020 (Schlick and Bernhart 2013). But these improvements in the use-stage will lead to additional environmental burden and financial efforts in the manufacturing stage of a car (Rödger et al. 2016). The production and provision of high-end batteries for electric cars cause environmental impacts and in the use-stage due to the additional weight as well (Helmers and Marx 2012; Frischknecht and Flury 2011). In terms of relative sustainability (each car seen in isolation) the sector is on a good way, but as soon as absolute sustainability (total impact of the car sector) is considered, a contradictory picture can be drawn. Increased global car sales diminish the specific savings per car due to larger population and higher prosperity around the world—the so-called rebound effect (as mentioned in Sect. 4.2). It can be concluded that a more sustainable car sector can only be achieved if the manufacturing and the EoL stage of the single car experience a similar interest of effectiveness compared to the use-stage of the product (Rödger et al. 2016).

To meet the increasing market demand for cars, automation technologies experienced a tremendous increase of interest (IFR 2013). Automated production is usually applied where it provides a possibility of capacity increase, improved productivity, reduction of manpower, reduction of repetitive and hazardous work tasks and higher product quality (Bellgran and Säfsten 2010). The car body production, as a good example due to its high degree of automation (>95%), consists usually of three separated production lines, which are producing the underbody, the body frame and the doors, fenders, etc. These lines are designed with the focus on two main indicators—the annual output and the required cycle time. In a typical body shop, around 300–500 parts are assembled by about 1.200 robots before sent to the paint shop (Galitsky and Worrell 2008). The degree of automation increased over the last years and the vehicles-produced-per-robot-ratio is declining, e.g. in Germany since 2007 from around 72 down to 62 in 2014 (Rödger et al. 2016). This trend is set to continue in the main robot markets (Japan, Rep. Korea, China, North America, Germany and Brazil), and an increase of industrial robots in operation of up to 23% until 2016 is expected (IFR 2013).

To illustrate the impact of automation, the direct energy consumption of robotic body shops compared to the entire production is shown in Fig. 4.2. Only 6% of the overall consumption is directly due to the body shop with all its different components. Of this, robots consume around 30%, the joining technologies 20% and the infrastructure (like ventilation, lighting and cooling) the remaining 50%.

The average CO_2 equivalent emissions over the life cycle of an industrial robot (with a payload up to 250 kg) used in Germany in a typical body-in-white line can be assumed with around 24–42 tons over up to seven years of operation (Drechsel et al. 2015; Dijkman et al. 2015). Klüger (2013) states that a typical body-shop robot consumes around 4.500 kWh of electricity per year. Based on the analysis from (Dijkman et al. 2015), it means that roughly 65% of the overall emissions are caused during the use-stage and additional 35% must be considered due to the production and end-of-life. It has to be mentioned that the direct energy consumption of the robot is highly dependent on the application and should be analysed case specific.

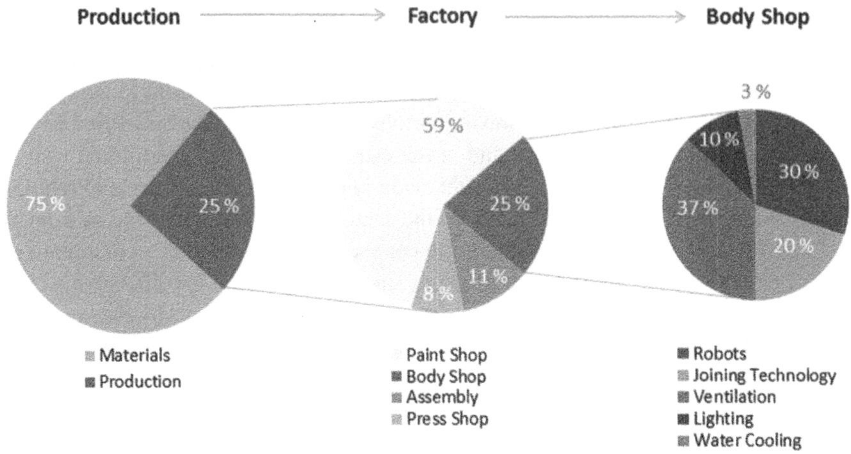

Fig. 4.2 Estimation of the share of energy consumers in a European car production from top down according to Heil et al. (2014) and Klüger (2013)

Another aspect is the additional initial investments for automation. A price of 20.000–35.000 € can be assumed for an industrial robot, although considerable discounts are often available, e.g. depending on the number of robots ordered. Considering an industry price for electricity of 0.08–0.12 Euro-cent per kWh (BDEW 2013) and an operating life of 6 years for a robot, total direct costs of around 27.000–38.000 € can be expected for the robot user. Considering the rise in robot units used per production facility expected in the future, increasing investment costs and increasing electricity expenses for the companies can be expected. Due to this fact, new methods for investment decisions are already applied in industry which consider the life cycle—total cost of ownership (TCO) or life cycle costing (LCC) (UNEP/SETAC 2011).

Social aspects are the last but not least pillar in sustainability. In automated production, social impacts are a challenging topic. The latest developments indicate that Social LCA (explained in Sect. 4.2) is used as a tool for corporate social responsibility (CSR) to screen an organization's supply chain for social hotspots or to support socially sustainable procurement (Norris and Revéret 2014). The planning of an automotive production line typically includes a number of stakeholders such as the original equipment manufacturer (OEM, car producer), the system integrator plus suppliers of application, software and equipment. Defining generically applicable social indicators for all stakeholders is challenging, since contexts are different for each stakeholder. The UNEP/SETAC methodological sheets do offer a great choice of indicators, but many of those are not suitable for the direct production as they capture impacts late in the impact pathway and do not offer enough detail to establish whether or not the current production is socially sustainable (Norris et al. 2013). To determine which type and composition of indicators may bring the most useful contribution to the planning of a socially sustainable industrial production, a characterization step has to be discussed as an option. Especially, the quantification of potential impacts due to physical and physiological exposures on social and psychological mid- and endpoints is not solved yet.

4.4 Sustainability-Related Decision-Making in Highly Automated Car Manufacturing

The planning process of the production is closely linked to the product. Everything starts with a company's strategy and the product concept around six years before the serial production of a car can start. Along this process, several quality gates must be passed. After around 2.5 years, the quality gate "design freeze" has been reached, where design and concept of the car are finished. Based on the product data and design, the layout of the production line and specific machine tools will be developed. From that point until the start of production (SOP), there are several decisions to be made which may have impact on the sustainability performance. But the question is how and where these decisions should be implemented in the most feasible way without compromising the established production planning process.

In general, the planning process is a top-down approach and therefore targets should be considered as early as possible to exert the greatest influence with limited financial effort (Bellgran and Säfsten 2010). The production line of a car must be capable of several derivatives (i.e. versions of a car model), heterogeneous processes and meet the expected sales figures (VDI 2008). To achieve these needs, the planning is done with Digital Factory Planning software (e.g. from Dassault Systèmes or Siemens). The planning process is usually divided into three phases which are explained below, including a description of impacts which each phase has on the three dimensions of sustainability:

1. Original Equipment Manufacturer (OEM) supplies a complete resource list, and system integrator (SI) elaborates a functioning model of the production line including investment costs
   Impacts:
   a. Economic: High, due to the predefined resource list and the layout of the production line.
   b. Environmental: A number of machinery and tools are defined that leads to—in terms of the life cycle perspective—a fixed amount of contributions to impacts already from the beginning. Although the use-stage is still dominating (as long as the carbon dioxide intensity of electricity is high), indirect effects like occupied area and occupied volume have been already decided upon as well and these are associated with additional contributions to impacts.
   c. Social: Only small number of decisions, e.g. how many workers should be considered in the different line.

2. Subsequent product changes (like joining sequences) are integrated by the SI and the model is adapted. The SI assures a full functional line in terms of annual output and cycle time.
   Impacts:
   a. Economic: Although a lot of important decisions regarding the manufacturability (e.g. detailed cycle times of each cell) or rather regarding the actual use of the line are made here, only small changes compared to the fixed initial costs from step one can be estimated. The expenses for direct energy consumption in automated manufacturing are very low, compared to the investment costs.
   b. Environmental: The environmental impacts are fairly high, because the use-stage of the line will be defined. However, the direct energy consumption of automation systems is roughly 50% of the overall consumption in a production line; therefore, the impact is limited as well.
   c. Social: Some impacts regarding noise and emissions have to be dealt with, due to the technology choice and design (e.g. gluing and welding have completely different emissions and noise levels).
3. New planning of the line with individual ideas. The provided machine operating times and non-productive times from the OEM Standard must be complied with. A new resource list, layout and cycle time diagram must be prepared incl. adjusted investment costs.
   Impact:
   a. Economic: High, because the system integrator can supply a completely new solution.
   b. Environmental: Fairly high, the system integrator can design the line in the most sustainable way regarding direct energy consumption. The machine tools and so forth, still carrying a lot of impact, and the indirect consumption like lighting and some parts of the ventilation depend on the infrastructure of the factory.
   c. Social: Fairly high, because emission levels and working environment can be enhanced.

It can be concluded that the existing planning process is capable of three sustainability assessments with small additional time effort under the condition that all relevant information can be provided by the planning software. Additional benefit would be that the planners could compare their ideas in all stages, implementing improvements at the same time and use these results to create competitive edge. Especially, the first planning phase where the main decisions are taken could benefit from implementing sustainability performance data like life cycle inventory data from EcoInvent® or thinkstep® (formerly known as PE-International).

The above-mentioned decisions are only viable along the production line planning. Other sustainability-related impacts are made additionally during the realization and ramp-up phase. Especially, the social dimension is very important along the realization phase, because the facilities are tested regarding compliance of certain thresholds

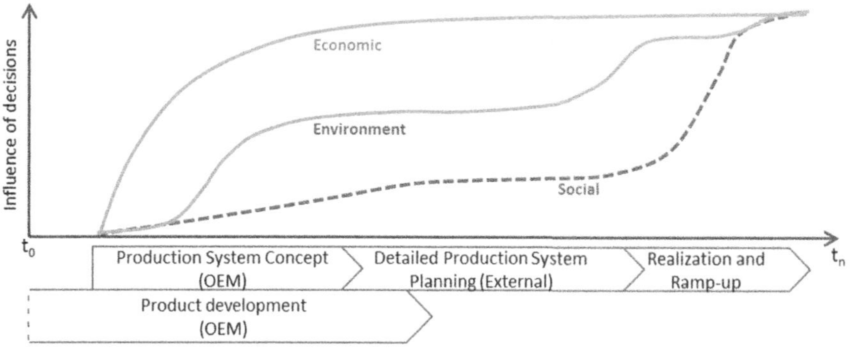

Fig. 4.3 Illustration of principle influence on sustainability aspects (economic, environment and social) exerted by the planners during the planning process of a production line for greenfield projects (Rödger et al. 2016)

like dust, noise and fumes. Figure 4.3 illustrates and summarizes the influence of the decisions on the three sustainability dimensions along the whole production planning process in a qualitative way.

The economic aspects are dealt with very early in the planning process compared to the environmental aspects. Also the basic layout, the joining sequence and therefore the number of robots have been designed early, which is represented by the increase of impacts at the end of the production system concept phase. Along the detailed planning, the direct energy consumption can be adjusted or even new concepts can be presented (illustrated by the second steep increase at the end of the second phase). The possibility to influence social aspects increases along the planning process, and during the realization and ramp-up phase all measurements are implemented and adjustments are made, illustrated by the steep increase.

It can be concluded that the first phase of the planning process is most important, and within the first six months it seems possible to enhance the sustainability profile of the production significantly. To support this, a method is needed, but no methodological framework exists so far which links the product emissions to the production. Therefore, a new approach (Rödger et al. 2016) describes how to combine absolute targets for the product and the allocation to the various production levels. These sub-targets should be used and met by the production line developers.

4.5 Framework for Sustainability Assessment of Highly Automated Production

Automated manufacturing is part of a complex system with many layers that ultimately contributes to fulfilling a demand from the market for a given functionality. This demand is satisfied by new products and therefore new production lines have to

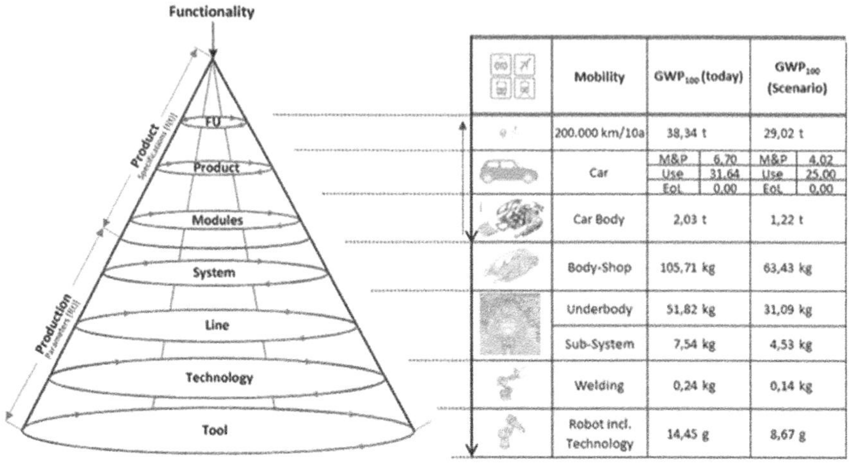

Fig. 4.4 Sustainability cone—framework of absolute goals for the functionality and allocation of specific targets to different levels in manufacturing (Rödger et al. 2016)

be developed. To meet this demand in a sustainable way, the life cycle impact of the product must be considered, specifications have to be passed, top down, from market demand (functionality) down to the resulting product, the components, the production system of these components, down to among others the automation solutions used. This is illustrated in the cone-shaped model in Fig. 4.4 and leads to a framework which combines absolute life cycle targets of products with the ones in production and is able to define sub-targets for the whole production chain and therefore avoid sub-optimization or problem shifting (Rödger et al. 2016).

Starting with the assumption that each of the levels will have its own life cycle and thus individual environmental and financial targets. To get comparable results, the system boundaries of the life cycle were adjusted slightly by including the research and development (R&D) as well as the design stage. This means in particular that, e.g., the price of a car includes the costs for R&D but most environmental assessments do not.

To understand the brief description above, a detailed real example of a car is given below. The customer asks for a new car that is able to provide a reach of 200.000 km over 10 years (the so-called Functional Unit in Life Cycle Assessment). The strategy department besides others sets absolute environmental and financial targets for this product and those are translated by the designers into base specifications of a car, e.g. middle-class version and fuel consumption. These assumptions predetermine more or less the shares of the manufacturing stage, use-stage and EoL stage in regard to the strategy goals and have to be at least fulfilled or even exceeded.

In the early stage of product development, specifications are set to fulfil the Functional Unit (FU) that actually influences the manufacturing system. Components of the product are developed, and further detailed specifications will evolve for the components and for how they will be manufactured in the production system (e.g.

layout of the body-shop line). According to, e.g., Heil et al. (2014), the material composition of the car body, the design, number of derivatives or even the location affect the layout of the system, the technology and the tools to choose. All this leads to specifications for the manufacturing technology, such as welding or cutting, and those need different machine tools (e.g. robots).

To explain the approach itself, the global warming potential over 100 years will be used as an exemplary impact category, but several other categories can be implemented in this approach to avoid sub-optimization between them, based on a middle-class car which emits 38.3 t CO_2 over the entire life cycle (see third column in Fig. 4.4) and reflects the median of the available simplified LCA results (as-is situation, see Sect. 4.3). If a successor has to be developed, the designer and developers have to consider and comply with various prospective circumstances. Assuming that the car manufacturing does not want to increase their total greenhouse gas emissions and comply with new directives (to avoid fines), the manufacturing emissions have to be reduced by 40% because of the increased number of sales until 2025 (in order to mitigate rebound effects) and fuel consumption by 21% to comply, e.g., the EU-directive 333/2014. This leads to about 8.3 t of CO_2 less along the life cycle compared to the predecessor, whereas manufacturing has to contribute with about 2.7 tons (1.7 t in material and 1 t in production). Assuming the same share between material and production intensity on the product level and similar energy intensity of the production steps (25% for body-in-white), the new car body is allowed to emit 1.2 t in total including material, infrastructure and indirect energy consumption. The body-in-white production line consists of three different lines, and it is assumed to consume around 40% in the as-is situation as well as in the scenario (63.4 kg). Including direct energy consumption for technologies and machine tools as well as infrastructure, indirect energy consumption for ventilation, heating and media supply, the production line is consequently allowed to a specific amount per underbody. The production system usually consists of three lines (e.g. Z1, Z2 and Z3), thus each line can get its specific target based on the superior level. Based on several widely used critical success factors, like number of joining equivalents, number of processes, occupied area, number of machinery, vertical range and empirical values, the targets can be determined and will be allocated. Assuming that Z2 and Z3 just get refurbished, Z1 (FU = one underbody) would receive most of the budget (e.g. 31.1 kg CO_2-eq). Those targets will be passed on to each subordinated level and therefore sub-optimization can be avoided. They can be used as targets for suppliers, e.g. as part of the tender. This approach can be applied in even more subordinated levels, and the carbon dioxide emission allowances for each robot in a specific cell can be determined (see Rödger et al. 2016).

To summarize, the above-mentioned amounts of carbon emissions always reflect the total amount of the life cycle of each level including the levels below. These specifications are determined always by the superior level and the levels beneath have to report their life cycle emissions to show if they have fulfilled or even exceed the targets. It can be concluded that by this approach the functionality and therefore the life cycle emission of the product can be allocated to the sub-levels to fulfil strategic targets of the product. To achieve this in a feasible way, these targets should be included in the already existing and well-established process of specifications for the internal departments, which are involved in the production line planning process. Additionally, the planning software must be enhanced with life cycle inventory data to enable the planners and designers to develop the line in the most sustainable way.

4.6 Conclusions

The prosperity and the population are increasing steadily in the main Asian and African countries, and therefore, more products are demanded and have to be produced. To limit the environmental impact, or at least to keep it at the same level, the only lever to generate the required substantial reduction is the factor technology, which directly relates to products and manufacturing.

There are several approaches available in the scientific world to assess sustainability in manufacturing, but two crucial elements to reach holistic solutions are often not covered—firstly, considering the life cycle of the product in strategic planning and secondly, deriving product-related sub-targets for the production system which contributes significantly to the environmental and financial targets.

Automated manufacturing will play a major role in maintaining or increasing competiveness of the companies globally. This implies that the sales of industrial robots will increase due to demand of capacity increase, improved productivity, reduction of manpower, reduction of repetitive and hazardous work tasks and higher product quality. However, this trend will also lead to some drawbacks like higher initial costs and environmental impacts in manufacturing.

As a response, the decision-making has to be improved by using already existing tools and software solutions and integrating specific targets for all levels in complex production systems. In order to achieve this, digital factory planning software should be enhanced with LCA and cost data to support, in particular, the crucial first phase of production line planning. In this way, the sustainability performance of the production line can be tracked and improved already from the early planning stages.

As a mean to set targets on each level of the product development process, the sustainability cone is introduced as a new framework. By using this framework, production line planners and others are able to break down life cycle emission targets of a product to subordinated levels, incl. the manufacturing. To implement resulting sub-targets in such a mature industry, we suggest to integrate them in already existing and used specifications sheets, which internal departments and external suppliers have to comply with.

There are several challenges ahead to integrate this framework in the decision-making process of such a mature industry. Firstly, the strategic department and product developers have to adopt the life cycle thinking approach and should be keen on integrating it into the planning procedure. Secondly, planning software must be enhanced with life cycle inventory data to enable planners and designers on the different levels to assess their ideas in a feasible way. Thirdly, the communication between the levels (e.g. between system integrator and robot supplier) should be enhanced strongly. This might be the biggest challenge, because the subordinated levels always have to reveal a lot of information about their products (e.g. material compositions of what they supply). However, such a requirement can evolve into providing a competitive edge for those suppliers who are willing to share their information.

Applying this framework in several industries is planned in order to verify assumptions and the approach.

Acknowledgements The research leading to these results has received funding from the European Community's Seventh Framework Program under grant agreement No. 609391 (AREUS).

References

Amrina E, Yusof SM (2011) Key performance indicators for sustainable manufacturing evaluation in automotive companies. In: 2011 IEEE international conference on industrial engineering and engineering management, pp 1093–1097

Baldwin JS, Allen PM, Winder B, Ridgway K (2005) Modelling manufacturing evolution: thoughts on sustainable industrial development. J Clean Prod 13(9):887–902

BDEW (2013) BDEW-Strompreisanalyse Mai 2013, Berlin

Bellgran M, Säfsten K (2010) Production development. Springer, London

Benoït Norris C, Revéret J-P (2014) Partial organization and social LCA development: the creation and expansion of an epistemic community. In: Muthu SS (ed) Social life cycle assessment, vol. 6, no. 7. Springer, Singapore, Heidelberg, New York, Hong Kong, pp 4200–4226

Benoït Norris C, Traverso M, Valdivia S, Vickery-Niederman G, Franze J, Ciroth A, Mazijn B, Aulisio D (2013) the methodological sheets for sub-categories in social life cycle assessment (S-LCA). UNEP

Bey N, Hauschild MZ, McAloone TC (2013) Drivers and barriers for implementation of environmental strategies in manufacturing companies. CIRP Ann Manuf Technol 62(1):43–46

Bjørn A, Hauschild MZ (2013) Absolute versus relative environmental sustainability: what can the cradle-to-cradle and eco-efficiency concepts learn from each other? Bjørn and Hauschild cradle to cradle versus eco-efficiency. J Ind Ecol 17(2):321–332

Borucke M, Moore D, Cranston G, Gracey K, Iha K, Larson J, Morales JC, Wackernagel M, Galli A (2013) Accounting for demand and supply of the biosphere's regenerative capacity: the National Footprint Accounts' underlying methodology and framework. Ecol Indic 24:518–533

BSI (2011) PAS 2050:2011 specification for the assessment of the life cycle greenhouse gas emissions of goods and services

Carson R (1962) Silent spring. Houghton Mifflin Company, Boston

Dijkman TJ, Rödger J-M, Bey N (2015) Managing the life cycle of production equipment : what does it matter ? In: LCM conference Bordeaux

Drechsel M, Bornschlegl M, Hackbarth D (2015) The value and necessity of using environmental direct comparisons as a simplified life cycle assessment method in industrial planning processes. In: Electric drives production conference & exhibition

Du C, Freire F, Dias L (2015) Overview of social life cycle assessment. In: Life cycle in practice

Duflou JR, Sutherland JW, Dornfeld D, Herrmann C, Jeswiet J, Kara S, Hauschild M, Kellens K (2012) Towards energy and resource efficient manufacturing: a processes and systems approach. CIRP Ann Manuf Technol 61(2):587–609

EFFRA (2013) Factories of the future. European Union, Brussels

EU JRC (2014) Guidance for the implementation of the EU Product Environmental Footprint (PEF) during the Environmental Footprint (EF) pilot phase, pp 1–55

EU (2014) No 443/2009 to define the modalities for reaching the 2020 target to reduce CO_2 emissions from new passenger cars, pp 15–21, Mar 2014
Finkbeiner M, Schau EM, Lehmann A, Traverso M (2010) Towards life cycle sustainability assessment. Sustainability 2(10):3309–3322
Frischknecht R, Flury K (2011) Life cycle assessment of electric mobility: answers and challenges—Zurich, April 6, 2011. Int J Life Cycle Assess 16(7):691–695
Galitsky C, Worrell E (2008) Energy efficiency improvement and cost saving opportunities for the vehicle assembly industry, U.S. Environmental Protection Agency, Berkeley
Garetti M, Taisch M (2012) Sustainable manufacturing: trends and research challenges. Prod Plan Control 23(2–3):83–104
Graedel T, Allenby BR (1995) Industrial ecology, vol 20, no 4. Prentice Hall
Hauschild MZ (2005) Assessing environmental impacts in a life-cycle perspective. Environ Sci Technol 39(4):81A–88A
Heil U, Paulus I, Schacht M (2014) Karosseriebautage Hamburg, pp 3–13
Helmers E, Marx P (2012) Electric cars: technical characteristics and environmental impacts. Environ Sci Eur 24(14):1–15
Herrmann C (2010) Ganzheitliches Life Cycle Management. Springer, Heidelberg, Dordrecht, London, New York
Hoekstra A (2010) The water footprint : water in the supply chain. Environmentalist (93):12–13
Hoekstra Y, Mekonnen MM (2012) The water footprint of humanity. Proc Natl Acad Sci 109(9):3232–3237
IFR (2013) Executive Summary 1. World Robotics 2013 Industrial Robots 2. World Robotics 2013 Service Robots 3. World Robotics 2013 Industrial Robots. World Robotics Report—Executive Summary, pp 10–21
ISO 14040 (2006) Environmental management—life cycle assessment—principles and framework
Jørgensen A, Hauschild MZ, Jørgensen MS (2010) Developing the social life cycle assessment: addressing issues of validity and usability. Dissertation, Feb 2010
Klüger P (2013) Future body-in-white concepts increased sustainability with reduced investment and life cycle costs
OECD/IEA (2012) Technology roadmap—fuel economy of road vehicles, Paris
Rockström J, Steffen W, Noone K (2009) Planetary boundaries: exploring the safe operating space for humanity (html). Ecol Soc 14(2)
Rödger J-M, Bey N, Alting L (2016) The sustainability cone—a holistic framework to integrate sustainability thinking into manufacturing. CIRP Ann Manuf Technol (submitted)
Schlick T, Bernhart W (2013) The development of e-mobility in global markets. In: IAA 2013
Schrettle S, Hinz A, Scherrer-Rathje M, Friedli T (2014) Turning sustainability into action: Explaining firms' sustainability efforts and their impact on firm performance. Int J Prod Econ 147:73–84
Steffen W, Richardson K, Rockström J, Cornell S, Fetzer I, Bennett E, Biggs R, Carpenter SR, de Wit CA, Folke C, Mace G, Persson LM, Veerabhadran R, Reyers B, Sörlin S (2015) Planetary boundaries: guiding human development on a changing planet, prep, Jan 2015
Thiede S (2012) Sustainable production. In: Life cycle engineering. Springer
Traverso M, Finkbeiner M, Jørgensen A, Schneider L (2012) Life cycle sustainability dashboard. J Ind Ecol 16(5):680–688
UNEP/SETAC (2011) Towards a life cycle sustainability assessment. UNEP/SETAC life cycle initiative
VDI (2008) Digitale Fabrik Grundlagen Digital factory Fundamentals VDI 4499. System
Warhurst A (2002) Sustainability indicators and sustainability performance management 43(43)
World Bank (2011a) data.worldbank.org. http://data.worldbank.org/indicator/NV.IND.MANF.ZS. Accessed 26 June 2014
World Bank (2011b) Manufacturing, value added (% of GDP). http://data.worldbank.org/indicato r/NV.IND.MANF.ZS

Chapter 5
Piloting Comprehensive Industrial Energy Efficiency Improvement in a European Rolling Stock Factory

Nils Weinert, Rafael Fink, Christian Mose, Friedrich Lupp, Florian Müller, Jan Fischer, Ingo Bernsdorf and Alessandro Cannata

5.1 Introduction

A major share of the primary energy consumed globally has to be accounted to manufacturing, as well as related emissions have to. Improvement of energy efficiency in factories is therefore a key driver to support the achievement of the European 20/20/20 goals. Hence, industry has to rethink current approaches about design and management of manufacturing systems to take a significant step towards energy-efficient factories.

However, the task is challenging, since factories are complex systems made of interacting elements such as people, production assets, material handling equipment, building service equipment. A holistic approach which, moving away from local optimization, considers the factory as a whole has been developed within the European Research project EMC^2-Factory.

To show the industrial applicability, the approach has been tested in a real case, a European rolling stock factory. For this factory, several measures to improve energy efficiency have been identified and were developed, of which the most relevant and impactful are presented in this book section.

5.2 Approach Adopted

The EMC^2-Factory project aims at developing new solutions for planning and operating eco-efficient factories, as well as using existing methods and tools combining them in a more effective way. An integrated view has been adopted in the project to prevent problem shifting due to the mentioned complex system interconnections,

N. Weinert (✉) · R. Fink · C. Mose · F. Lupp · F. Müller · J. Fischer · I. Bernsdorf · A. Cannata
Siemens AG Corporate Technology, Munich, Berlin, Germany
e-mail: nils.weinert@siemens.com

S. Thiede and C. Herrmann (eds.), *Eco-Factories of the Future*, Sustainable Production, Life Cycle Engineering and Management,
https://doi.org/10.1007/978-3-319-93730-4_5

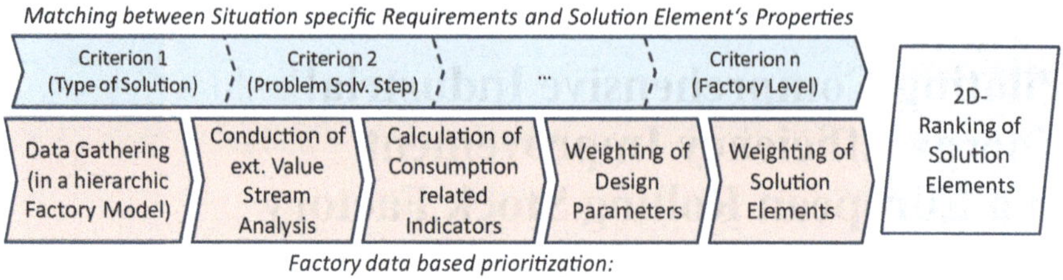

Fig. 5.1 Prioritization of Solution Elements in the solution finding process (based on Fischer et al. 2014)

e.g. from production to technical building services. This integrated view fully considers all the components in a factory and their reciprocal influence on each other. Solutions (both developed within EMC^2-Factory and available externally) for this variety of fields are very diverse and end up in a high number of alternatives that the decision maker can choose from. Moreover, since factories are significantly different for every specific case, this complicates the decision-making for energy efficiency improvement in the individual industrial case.

Hence, a methodology for systematically and objectively prioritizing these possible improvement measures (here called "Solution Elements") for concrete factories is required that was developed within the scope of the project. This "Solution Finding Methodology", of which details were presented in (Fischer et al. 2014), allows the structured and objective evaluation of all kinds of possible Solution Elements in a widely automated way. In order to allow an easy applicability, it was implemented in a prototype software tool.

The methodology is substructured in two phases (see Fig. 5.1):

- Phase 1: A (quick check) prioritization phase where the properties of the Solution Elements are matched with the current demands and user preferences coming from the current perspective of the factory and the step in the optimization process;
- Phase 2: A detailed prioritization phase where the Solution Elements are weighted based on a multi-step selection process, using actual factory data as derived in an extended value stream analysis (and then further analysed).

The separation in these two phases allows the exploitation of both, the quickly evaluable but partially subjective characteristics of Solution Elements on the one hand and the more detailed and objective evaluation based on factory data on the other hand.

In order to allow the Solution Element's evaluation by the described process phases, they were formalized and stored in a machine-interpretable form. This standardized form was defined in the Solution Elements Standard Sheet, allowing the description of the Solution Elements among a defined set of seven properties and their possible property values as, e.g.:

- The type of Solution Element (e.g. technical solution, framework, process description, software tool);

- The applicability during problem-solving steps (e.g. data gathering phase, analysis phase, implementation phase);
- The type of goals that are targeted (e.g. energy efficiency, quality improvement, cost efficiency);
- The applicability at a factory level (e.g. site level, building level, area level, machine level)
- The maturity level (e.g. under research and development, in use in some areas, widely used).

In the quick check phase of the solution finding process (SFP), these properties are evaluated in a matching process, comparing the user's preferences with the values of the Solution Element's properties. In order to account for the sometimes fuzzy relationship of Solution Elements towards some property, assignments can be specified by numerical values, hence representing a fuzzy-like connection.

The second phase of the SFP makes use of the fact that Solution Elements have different strengths of correlation with certain reoccurring areas of improvement in factories (Step 5 in Phase 2 in Fig. 5.1)—here called Design Parameters. For example, the Kanban methodology mainly targets the production control mechanism in a factory while the use of light-emitting diodes (LEDs) mainly affects the lighting technology that is used. The numerical correlation between Solution Elements and the Design Parameters is described by a property in the Solution Elements Standard Sheet, using Likert scales in order to translate a verbal strength of correlation into an approximated numerical one. In total, 25 Design Parameters are used that were previously derived (during the development of the methodology) in a systematic selection process.

In order to make use of this correlation, in Step 4 the relevance of the Design Parameters was weighted by a set of indicators that have a direct influence on the financial expenses and the energetic consumption of a value stream. The weighting is conducted by a matrix multiplication that maps these indicators (e.g. the process time, the set-up time, the energetic idle load or the bound capital of a machine) to the Design Parameters (Step 4). These indicators were previously derived via formula from the total energetic and financial consumption of a value stream that were derived in an extended value stream analysis EEVSA (Step 2 and Step 3).

This extended value stream analysis puts a special focus on financial and energetic expenditures related to value streams. In addition, all levels of peripheral systems in the actual factory buildings are considered as their consumptions are dependently allocated to the individual processes of the value stream (according to Posselt et al. 2014).

The required data in order to conduct the extended value stream analysis were collected in a structured process (during the first process step) and transferred in a digital factory model, describing the physical elements of a factory that have a relevant energetic or financial impact.

All steps of the described process are supported by the methodology developed within the project (e.g. the generic class structure of the factory model, the allocation formula for the extended value stream analysis or the correlation matrix between Design Parameters and consumption-related indicators).

5.3 Pilot Case Selected and Motivation

The selected pilot case is a Siemens AG rolling stock manufacturing site located in Vienna. It produces rail vehicles of stainless steel or of aluminium for metros, coaches and light rail. In the factory, several processes are performed, from machining and grinding to joining, coating and final assembly, in an area broader than 120 000 m^2. Before this research activity, the site already addressed the topic of energy and resource efficiency with different projects on building renovation (insulation, ventilation systems, etc.), a new surface treatment centre and product design improvement from life cycle perspective (e.g. lighter materials to consume less energy during the use phase of the trains, use of recyclable material); however, the holistic perspective adopted in this project was not yet part of the considerations.

In order to show the impact of the developed approach on energy efficiency, the pilot has been focused on a single building of the site. In this building, the main train components such as undercarriages, roofs and sides of a train car are manually grinded and manually and/or automatically welded. Moreover, in the same building two machining centres are available for pre- and post-welding processes.

The main processes are the following, described in a simplified way (different products have different process chains):

- Milling (machining): It is carried out by two large machining centres. Machining is in this case usually needed to prepare the large components for the welding process.
- Welding: Large components are moved with cranes to the welding stations where different automatic and/or manual welding processes (GMAW) are performed.
- Grinding: This operation is performed to deburr welding seams for quality issues.
- Milling: Depending on the product and part, the component may be needed to be moved again to the machining centre (e.g. to prepare the next welding processes).

Some relevant features of the production environment in the selected building are worth to highlight:

- There is a high mix of automatic machines and manual work.
- Material flow within the building is very complex and product-dependent.
- Components are difficult to be moved within the building; usually, two cranes are simultaneously required to move long components. Hence, the availability of cranes is sometimes an issue limiting proper material flow.
- Welding produces gases which go into the building. Sometimes, windows are opened to dissolve welding gases into the open air.
- Workers can significantly influence energy consumption by controlling technical building services (lighting and ventilation system) and by using production equip-

ment (e.g. detection of leakages when using compressed air, rules for switching equipment on/off).

With an area of ca. 5000 m^2, the building was chosen for three reasons:

- Representative: It includes typical manufacturing processes of train production.
- Relevant: It shows complex material flow and interaction among processes.
- Challenging: It was recently improved from energy perspective such as air heat exchangers, building insulation and air doors. Hence, further improvement requires going beyond the state-of-the-art technologies and approaches.

5.4 Selection of Energy Efficiency Improving Measures

For the investigated building and the affected part of the metro car's value stream, the solution finding process was applied as described in the previous chapter. First, the as-is analysis has been performed, gathering both energy- and production-related information.

Based on one-time measurements, several shop floor visits and available production data, the required information to set up the hierarchic factory model and to derive the extended energy value stream could be acquired. The total energy consumption of the building (visualized as Sankey diagram in Fig. 5.2) indicated high consumption of peripheral hardware like lighting, heating and ventilation among others. It needs to be highlighted that although ventilation and exhaust have been distributed proportionally (space-/time-related) among grinding machining and welding in order to represent the actual current situation in the building, the main cause for the need of a ventilation system requirement is due to the welding process.

Using the results of the EEVSA, the solution finding process was performed with several iteration loops. Each iteration loop was defined with different user preferences according to the momentary step in the problem-solving process and the related focus of application (e.g. analysis vs. solution phase, process vs. process chain perspective). Figure 5.3 depicts the results of one exemplary iteration loop, targeting the analysis and data gathering phase and the value stream process utilizing the biggest electrical consumer.

After several iterations, the solution finder indicated the biggest potential for energy management, production planning, alternative welding technologies and layout.

5.4.1 Energy Transparency and Management System

One of the commonly adopted principles in engineering is that improvement has to be founded on exact (more or less) data; thus, knowledge gained from transparency is often seen as a prerequisite for improvement. Following this guideline, certain

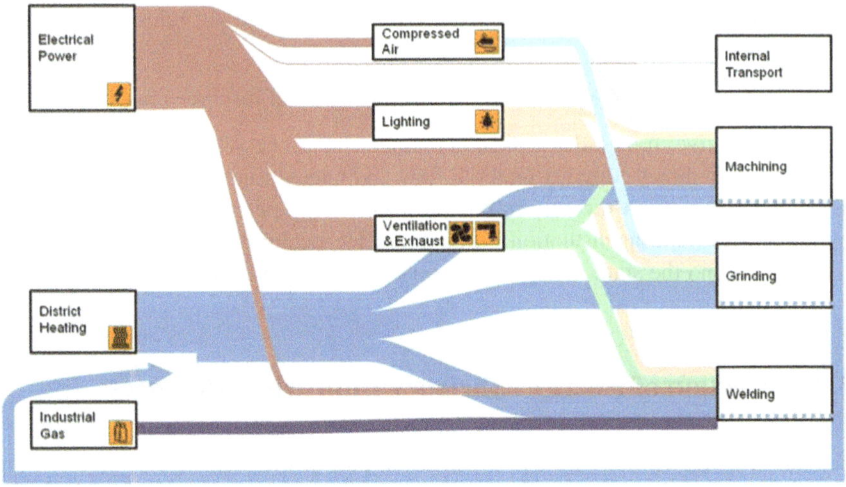

Fig. 5.2 Representation of energy Sankey diagram performed for the pilot case

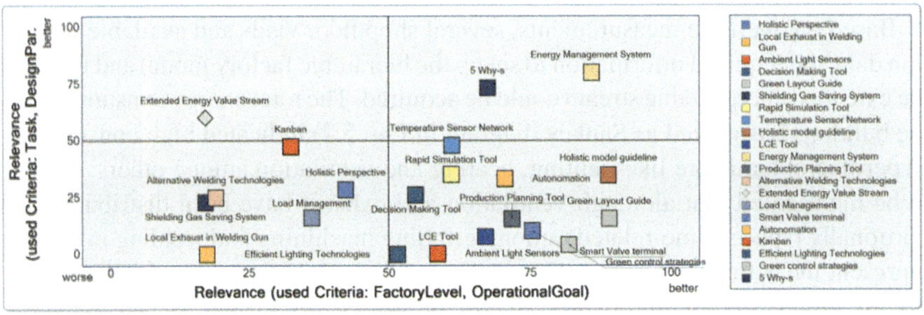

Fig. 5.3 Results of exemplary iteration cycle of solution finder

relevant production entities have been energetically measured using one-time measurements as a first step. The gained insights on energy consumption and distribution were used as input for applying the solution finder (Weinert et al. 2013).

One-time measurements are sufficient for providing a basic knowledge on the consumption in a steady application, but not in a dynamic environment like present in production, where continuous monitoring and analysing are required to detect shortages. Beyond developing the transparency required, actual management of energy consumptions becomes possible, leading to the decision to implement an energy management system (EMS) for the investigated building in the given case. The system consists of several measuring points for electrical energy and pressurized air and was designed for easily integrating further measured variables like heating or water supplies. As general for EMS Applications, an on-site display was implemented for providing direct feedback to shop floor or maintenance personnel (see Fig. 5.4).

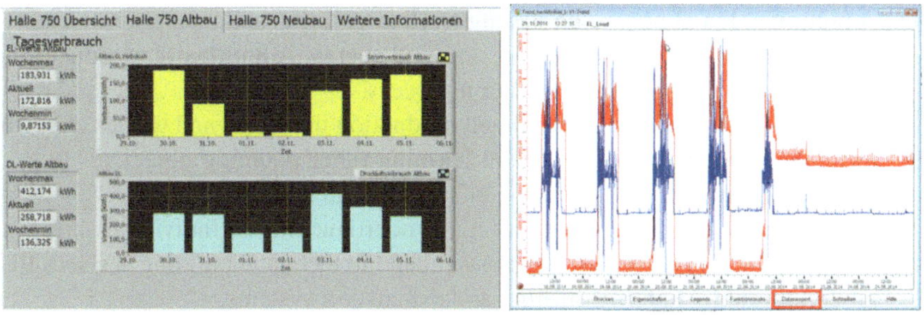

Fig. 5.4 Exemplary visualizations of energy consumed at on-site display

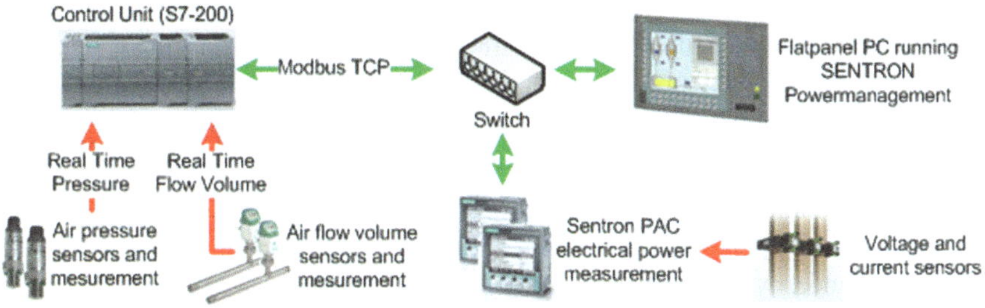

Fig. 5.5 Energy management system—concept

Measured consumptions are automatically monitored in real time on an industrial PC, generating signals for control measures like machine standby opportunities or air leakage detection. The overall system architecture (see Fig. 5.5) is designed for later enhancements, e.g. automated interventions in those cases, or for being integrated into an overall factory-wide energy management system as required, e.g. by ISO 50001, or even for future integration in smart grid environments.

5.4.2 Production Planning

Depending on the range and the complexity of the goods produced, production planning can be a difficult task. Presumably that is one of the reasons why most contributions from scientific literature and scheduling principles applied in practice just focus on single economic aspects like the adherence to delivery dates or the minimization of the make-span. Yet, production planning is a highly multi-objective task, in general. If, for instance, production planning does not explicitly account for potentials to improve energy efficiency, these potentials will probably not be tapped in the production execution phase, either.

In the following, there are therefore provided some levers related to energy efficiency that can be integrated into the production planning process:

(1) A minimization of non-value-adding work (e.g. set-up and transportation activities) helps to avoid the wasting of time and energy. Additionally, it reduces the logistic complexity of the production process.
(2) An appropriate choice of the utilized production machines and modes can help to reduce the energy costs without harming other planning objectives (e.g. if alternative machines are available or velocity can be regulated).
(3) A smart production organization (e.g. synchronization of individual shift systems) can help to reduce energy costs related to technical building services like air conditioning, lighting, heating and exhaustion systems.
(4) Load balancing and the minimization of peak loads can contribute to significantly lower energy costs if the peak load is charged and the production process is very energy intensive.
(5) The anticipation of applied control strategies (which, e.g. determine if and when machines are switched off during idle periods) can lead to a smoother course of production and therefore contributes to improve the energy efficiency.

Based on the circumstances that we found at the industrial pilot case, a multi-criteria production planning software tool was developed and implemented within the EMC^2-Factory project that addresses levers (1)–(3) besides the original economic objective. Hence, by using our software, production planners can now simultaneously optimize the production process with respect to different KPIs like the adherence to delivery dates, the needed transportation and set-up effort or the overall energy consumption. Trade-offs between these KPIs can be quantified by calculating and evaluating not just one single solution but a set of alternative non-dominated schedules. The algorithmic approach relies on the framework for multi-start priority rule-based schedule construction heuristics that we introduced in (Fink 2013). Therefore, in Fig. 5.6, we only summarize how we transferred that framework into the algorithm that we implemented in the software.

Levers (1) and (2) are directly addressed via the stochastic multi-criteria priority rules that determine the planning sequence of the orders and the choice of the machines and modes. Lever (3) can be integrated by simulating different shift scenarios independently from each other.

5.4.3 Evaluation of Process Energy

In a factory, the total energy consumption is significantly higher than the actual energy required for the main, value-adding manufacturing steps of a product. Besides, a complete process chain includes several pre- and post-processes consuming energy themselves, e.g. milling the edge of a work piece as preparation for a joining process. Furthermore, a large variety of secondary energy consumers is required, including maintenance and transport operations and technical building services (TBS).

All of those pre-, post- and secondary processes can be understood as additional energy consumers for enabling the actual value-adding manufacturing steps. Con-

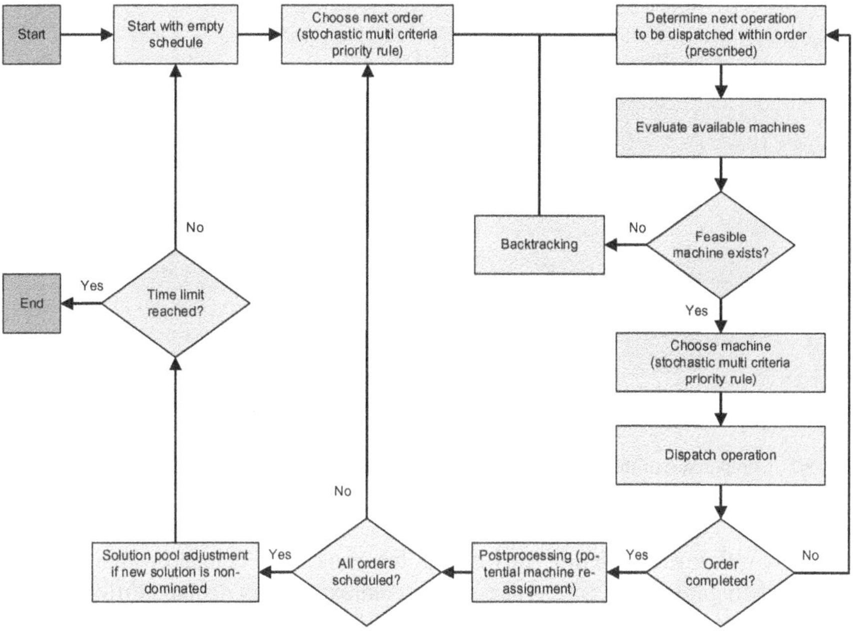

Fig. 5.6 Algorithmic concept of the production planning tool for the pilot case

sequently, in the design of a process chain it is not sufficient to focus on the main steps, but rather to investigate energetic demands holistically to achieve an overall optimum.

For doing so, process designers need the option to compare alternative process chains by providing an indicator, including the energy demand of all relevant primary and secondary consumers. This key element has been developed as a process-dependent normative energy indicator. Focusing on welding processes as the prevalent manufacturing process in the railcar industry, the specific energy demand per welded length is used. The indicator was defined as the accumulated energy over all considered process states of the manufacturing equipment and the whole production sequence, divided by the total length of weld created in that time resulting in the units [kWh/m] (Mose and Weinert 2013, 2015). Dominant factors influencing the indicator are found within the areas of:

- A particular process technology (e.g. GMAW or FSW);
- Particular process parameters for the observed work piece;
- Auxiliary systems implemented in the observed manufacturing equipment;
- Amount of non-productive time and associated energy demand on the equipment;
- Organizational influence by equipment's productive and non-productive time ratio;
- Organizational influence through the load of the equipment.

The approach introduced was applied in the Siemens plant to investigate alternative welding processes. Essential results are that major welding processes, like gas

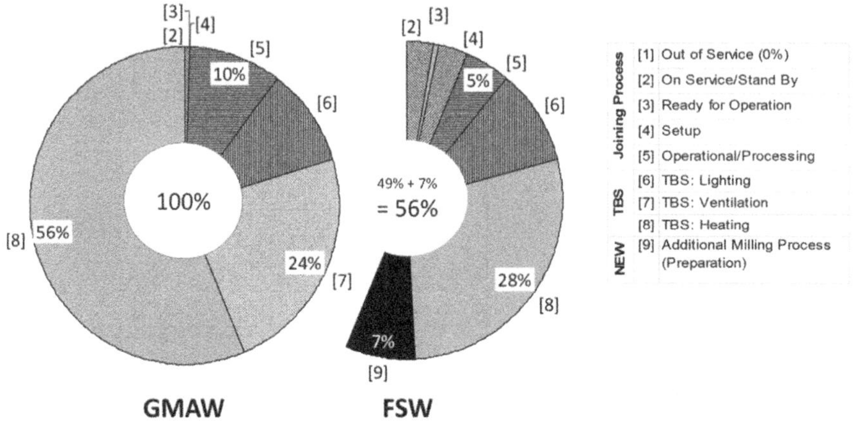

Fig. 5.7 Comparison of the differences in process chains for GMAW and FSW as core processes

metal arc welding and friction stir welding, have an energy demand within the same magnitude regarding the core processes. In contrast, energy demand differs significantly for secondary consumptions—mainly caused by different needs for exhaust systems—emphasizing the need to explicitly consider these in process chain design. However, a short-term change of process chains for existing products is difficult to realize due to the reasons like approved designs and processes, but will be considered for future product generations.

Different joining technologies, GMAW and FSW, were compared for similar joints. The particular energy demands are split into certain shares of demand according to different machine states. Technical building service (TBS) demands and addition processes were added as required to result in the similar outcome. All shares are considered as energy demands per welded length. The overall energetic balance of the process chains shows the result depicted in Fig. 5.7, with the main differences regarding energy demand being:

- Energy demand for the core process reduced to half ("FSW—Operation").
- Ventilation system is not required for FSW in difference to the GMAW process (no critical emissions).
- Massive reduction in heating demand for FSW being caused by not requiring the ventilation system compared to GMAW case (heat is not removed from the building through the ventilation system, and less fresh air has to be heated up for cold outside conditions).
- FSW process requires a higher surface quality than the GMAW process; therefore, an additional milling process is required as a joint preparation is necessary.

These single energy savings do add up to a 50% advantage for the FSW technology. As the surface requirements at the component's edges are higher for FSW, a share of this advantage is compensated by an additional milling process that is required as a preparational step for the FSW. Adding a process to the process chain in this case has a relatively small disadvantage compared to many other cases. The reason for this is that the equipment used for FSW is basically a retrofitted milling machine. Therefore, the milling process can be realized without additional handling and transportation of the work pieces, which would cause the major share of additional processing times and costs. Also, additional non-productive times can be kept very small. Of course, the additional tool for milling is required and the clamping fixture needs to be designed to support these two processes that are being realized on the machine just after each other. What was accomplished from the analysis done for this case was that considering the appropriate system borders is extremely important for any kind of analysis, even more when energy plays a major role. Of course, it is possible to optimize single manufacturing processes from an energetic prospective, but the major impacts that were discovered within this research showed that starting with a wider scope brought the main advantage. On the one hand, the indirect demands of energy and non-productive times both were identified to cause the major share of energy demand. On the other hand, it turned out that the consideration of the overall process chain is important. Calculating the same case with doing the milling on a different machine could add so much non-productive times and additional standby demand for milling machines to the calculation that there might not be an energetic advantage left.

5.4.4 *Process Energy from the Plant Perspective*

5.4.4.1 Plant Dilemma

Enhancing the approach further, it has to be considered that optimizing the manufacturing phase of an existing and of a new product cannot be investigated in an isolated manner. It has to consider the context of a manufacturing plant, where typically several products or product variants are manufactured, and therefore manufacturing phases meet. Focusing on the manufacturing phase, Fig. 5.8 displays process chains of three products with their processes of that particular phase lined up horizontally. Each workstation within a chain covers certain process steps. These can be understood as the implementation of particular manufacturing technologies, which are required at a certain point within one or more process chains for manufacturing. A plant then typically covers a certain segment of process chains, usually including several workstations. Process steps not covered by the plant can be understood as either outsourced processes, respectively, purchased services or components.

The economic optimization nowadays has led to a production planning behaviour of highly loading manufacturing stations with the purpose of lowering the manufacturing costs for optimization. Therefore, on manufacturing equipment it is common practice to use the equipment for processing a product mix to achieve the maximum equipment utilization. Changing process technologies in most cases requires equipment to be replaced. Introducing the replacement as a modification for one solely product—like it could be suggested as the first step and outcome of the comparison

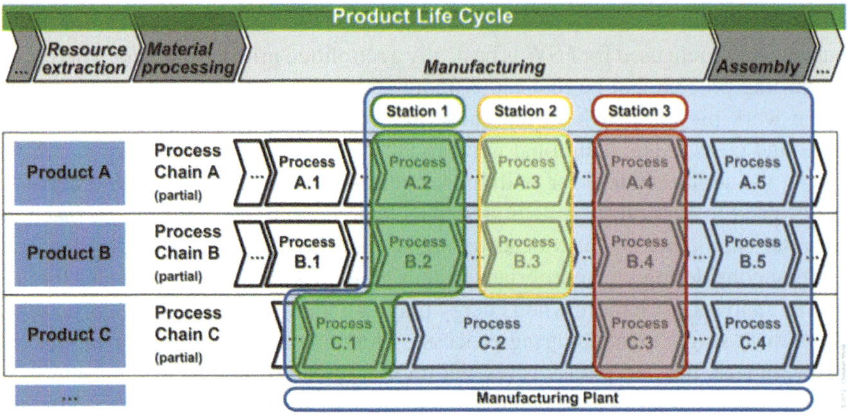

Fig. 5.8 Plant perspective on products' process chains and manufacturing station

of welding technologies—will have an economical impact not only on the modified production process but also on the factory's remaining product mix. This is caused by increased capacities which lead to a lower utilization of the workstations, and by required investments for new process technology. These effects result in higher manufacturing costs and make ecological-induced process adaptations both expensive and risky. Investing in environmental benign technologies therefore becomes economical malicious in multiple way and can be seen as a dilemma.

5.4.4.2 Sustainability Technology Roadmap

Although designing more energy-efficient process chains is possible by using the presented methodology, a great challenge is the implementation when facing today's predominantly economically driven decision-making in production planning. Single measures for raising the production sustainability or lowering a factory's CO_2 footprint in most cases are still economically disadvantageous. Today's predominant investment time frames are hard to be combined with CO_2 reductions requiring long amortization times. Targeted payoff periods which are derived from the pressure for rationalization do not leave much space for investing in sustainability improvements at the same time.

However, environmental performance of products and deriving from there to companies and factories increases in society's perception. With a growing demand for more transparency on the one hand, and environmentally consciousness on the other hand, sustainability becomes a competitive advantage. It is evident that in the long term sustainable production will become a necessity but also a matter of course for manufacturing companies.

As many of the identified more sustainable solutions are already economical bearable—just not in short term—it is necessary to find a way for harmonizing economical and ecological objectives. The required harmonization, as started in product development for the use phase (see "Eco Care Matrix" in Rohrmus et al. 2011), has to be started on factory level, based on the design and selection of process chains. As a tool for the systematic development of one's current manufacturing competences and technologies towards sustainable production, alternative process chains of different products have to be implemented stepwise, following a sustainability technology roadmap. When defining this roadmap, innovation and investment cycles have to be considered.

Following a roadmap-based approach fosters an adjusted planning of investment and innovation cycles. Investments for new technologies have to be adjusted with times of technology readiness in a meaningful way, what in an individual case could mean to adapt an existing process chain later than reasonable from an energy consumption perspective. To smoothen the transition towards a more energy- and resource-efficient production on the other hand, combined investments for production technology need to be considered.

A practical example for this idea is investing in a milling centre for modernization and extension of capacities of manufacturing processes, which contains an optional extension for friction stir welding (FSW), as compared to milling congener joining technology. The milling centre will be used for a period comparably long to the average production period of a product, and thus the existence of a particular process chain. In this example, the FSW ability will cause a higher invest at the time of purchase, although the process is not being implemented at that time. However, by providing the FSW option, this energy-efficient technology can be implemented for joining purposes later (following the sustainability roadmap for manufacturing) and then—when being implemented—with a significantly lower invest as if purchasing a FSW single-purpose machine at that point in time. Overall, this example illustrates the strategy for a smoothened transition from the current to a future technology and competence state of a factory, respectively company.

5.4.5 Extension Towards Factory Planning

Approaches in the field of factory planning differ in their adaptability to cases and the related degree of standardization. As demonstrated before based on factory optimization measures, challenges for green factories—factories both, environmental conscious and economically sound—cannot be addressed by solely adding environmental aspects to existing planning approaches, because interactions between factory elements are weakly addressed there. Moreover, every factory has its own drivers for energy consumption (e.g. production processes). This requires a planning which is unique for every factory planned and allows addressing its specific energy drivers.

The environmental performance is under constant cause-effect relations (between internal elements, external elements and Green Factory elements, see Fig. 5.9). Thus, green performance has to be integrated into a comprehensive concept rather than being an add-on to established approaches. Factory elements cannot be separated

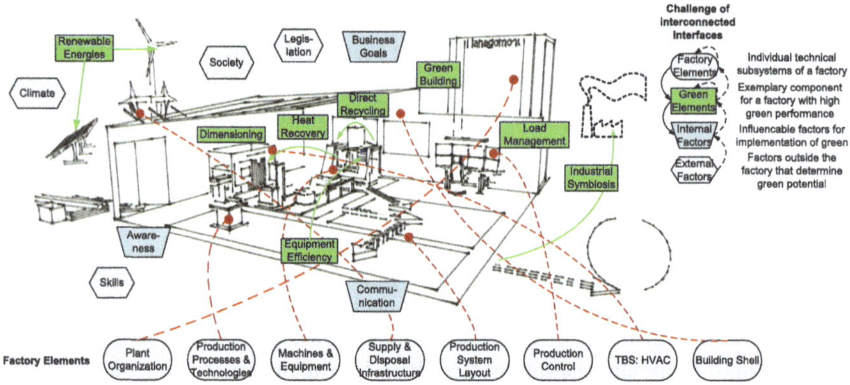

Fig. 5.9 Interconnected elements and factors for the planning of green factories (Müller 2015)

from each other; they compose the system factory and need to be specified as components that represent a high green performance. Internal (e.g. awareness or communication) and external factors (e.g. climate, skills or society) interact continuously with factory elements and green elements. For an optimal performance of the integrated factory, those interconnections have to be addressed.

The factory elements are defined as follows (Müller 2015):

- Organization: The organization sets the strategic and operation boundaries, requirements and structure for resource.
- Production Processes and Technologies: Several technologies that are used to fulfil a process function (e.g. GMAW or FSW for joining).
- Machines and Equipment: Equipment executes the production process or supports their execution (first periphery) which consumes energy.
- Supply and Disposal infrastructure: It connects all factory elements with each other to transport resources to the place where they are needed.
- Production System Layout: The whole value creation process takes place and is determined through material flow and storage. The production system is efficiently organized in terms of surface, and it serves as the area where production environment conditions interact with the equipment.
- Production Control: The way and infrastructure to process information and to intervene in a regulatory capacity in the production process predefines the load profile.
- Building—HVAC and Shell: Functions that ensure a regulated production environment and that separate production from the surrounding natural environmental influences.

An overall factory planning concept was developed that consists of five items (see Fig. 5.10): it is made of a philosophy to deduct the Green Factory vision (1) and align it with the strategic constituent parts towards green, the setting of targets and prioritization with KPIs (2) and a procedure (3) to manage the planning which consists

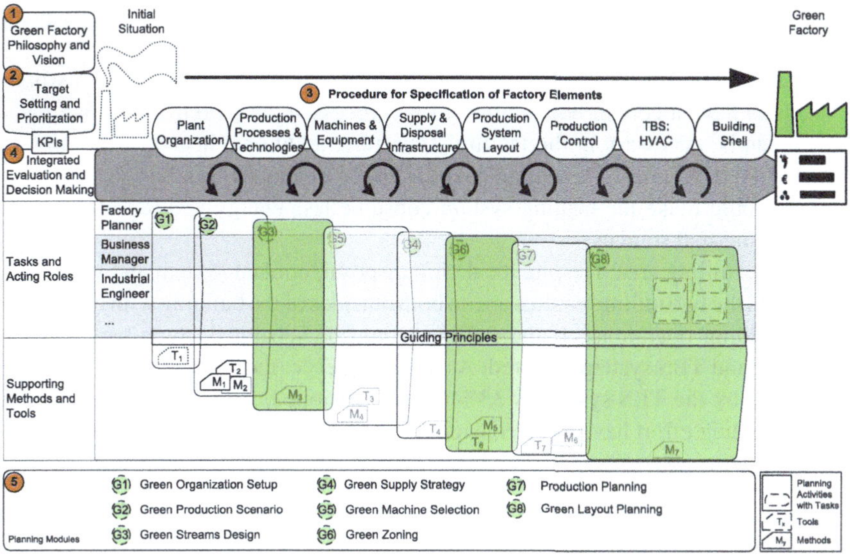

Fig. 5.10 Conceptual framework of Green Factory Planning (Müller 2015)

of an integrated evaluation and decision-making (4) to assess the specifications from planning modules (5). The coherence of items is considered in the concept of Green Factory Planning.

Overall, the factory element "production system layout" is of special interest for an energy-efficient design of buildings. A production system is the surface organization and the area where production environment conditions (e.g. illumination, temperature) interact with the equipment. While general planning guidelines like minimal floor load zones are state of the art in planning, the planning of elements like layout, TBS and disposal infrastructure across interfaces proofs to enforce communication, collaboration and a faster decision-making. Additional multidisciplinary experts provide knowledge and analytical methods together if the planning activity is properly designed.

For the pilot case, Green Factory Planning was tested by application of the planning module green layout planning. As a supporting method, a workshop-based green layout planning was developed to position machines and equipment with environmental aspects in mind (for details see Müller et al. 2013). Concerning the pilot case, three alternative layouts were developed that can be beneficial:

- Separation of welding and machining: Welding and machining of metal parts require filtering and exhaust ventilation of a different kind. If welding and machining operations would be locally separated smaller and better, designed TBS for one specific function could be used. A local decentralized exhaust system would be another option to avoid a central TBS system completely, but security and process

barriers caused by electrical conductibility of carbodies make this an unrealistic option.
- Isolated preheating zone for carbody: Because of quality reasons, the carbody needs to have a certain temperature to be processed, so a preheating time up to 24 h can be necessary in the facility. Therefore, additional heat will be supplied through the radiators. Preheating next to surface treatment to use waste heat from there would make the heating system consume less energy and reuse otherwise wasted thermal sources.
- Discrete milling area: Despite local chip disposal, coolant system and workplace lighting, the two milling centres are on both outsides of the building with machining and welding operations. If the milling would be separated, there would be no exhaust and TBS system required. As a consequence, the volume that needs to be serviced by the TBS system is 33% overdimensioned. Additional transport way and handling effort have to be considered for the production planning.

Since cost for a building and TBS change is disproportionately high, the implications have to be considered in the future building (re-)planning of the site. Nevertheless, they are general principles whose implications from environmental side can be easily identified in the layer evaluation approach (for details see Müller et al. 2013) for assessment.

Overall, the pilot case highlighted several insights to be considered for the application of Green Factory Planning in an industrial environment; for details, see Müller (2015):

- This work was based on the assumption that a more collaborative and communicative planning would be able to provide better performance of a developed factory. As demonstrated, the consideration of the complete system composed of production machines, processes, TBS and building shell enables better decision-making in reality. For example, the selection of production equipment solely on the energy intensity of the welding process would have had little impact, especially in comparison with addressing the substitution of welding technologies to avoid fumes at all.
- A further novelty of developed factory planning is the procedural description of planning modules like green layout planning, which guide experts towards considering green aspects in planning.
- Each case for environmental methods has unique features and characteristics. It is sufficient to apply a case-specific procedure that has methodical support for prioritization, because the created green factories have a better green performance than standard approaches. In the regarded pilot case, TCO and GHG emissions do not vary substantially if the additional building walls for spatial separation of welding and machining are not considered.
- The factory element production system layout does not cause energy consumption on its own, but it is a very intuitive and figurative base for factory planning. The layout has an integrative character since several technical elements interact on its base. Thus, planning can be improved with appropriate methods and tools, such as the layer visualization, if the layout is targeted.

5.5 Conclusions

The proposed approach started from creating transparency with an energy management system and covered different areas normally assessed only for production-related performance (production planning, substitution of processes and layout reconfiguration). The solutions highlighted from the approach are highly interdisciplinary, and they address the challenge of energy efficiency from different perspectives.

Some main remarks emerged from testing the approach:

1. Transparency is always the key starting point to initiate any energy efficiency program. Although energy is being monitored at factory level, more fine-grained data gathering is required in order to identify measures.
2. Energy efficiency should always be addressed comprehensively and considering case-specific features of a factory. To fully address energy efficiency, the complete system made of production machines, processes, TBS and building shell has to be analysed, thoroughly considering the interaction of all main components. For example, local optimization, such as the improvement of the direct energy requirement for welding process, can lead to little impact in comparison with addressing the reduction of the energy consumed by auxiliary systems for fume exhaust or by substituting process technologies to avoid fumes at all.
3. Although countless measures for energy efficiency improvement are available, each case has specific features and issues that can be almost unique. An approach that helps to structure the problem and focus on the most impactful measures is a promising approach to extend and facilitate decision-making when implementing energy efficiency measures in the industrial domain, either during factory planning or operation.
4. Financial return on investments on energy saving projects is still a transversal barrier. If on the one side economical impact is increased by more effective technologies and volatile energy prices, on the other side the effort in developing suitable solutions to finance energy saving projects should be continued.

References

Fink R (2013) A priority rule based multi criteria scheduling framework for energy efficiency aware detailed production planning. In: Proceedings of the 39th IEEE IECON. Vienna, pp 7508–7512

Fischer J, Weinert N, Herrmann C (2014) Method for selecting improvement measures for discrete production environments using an extended energy value stream model. Presented at the 12th global conference on sustainable manufacturing. Johor Bahru, Malaysia

Mose C, Weinert N (2013) Evaluation of process chains for an overall optimization of manufacturing energy efficiency. In: Advances in sustainable and competitive manufacturing systems. FAIM, pp 1639–1651

Mose C, Weinert N (2015) Process and process chain evaluation for an overall optimization of energy efficiency in manufacturing—the welding case. Robot Comput Integr Manuf 34:44–52

Müller F (2015) Modular planning concept for green factories. Doctoral dissertation submitted to the Technical University of Braunschweig. Braunschweig, Germany

Müller F, Cannata A, Stahl B, Taisch M, Thiede S, Herrmann C (2013) Green Factory Planning—framework and modules for a flexible approach. In: APMS 2013, Volume 414 of IFIP advances in information and communication technology. Springer, pp 167–174

Posselt G, Fischer J, Heinemann T, Thiede S, Alvandi S, Weinert N et al (2014) Extending energy value stream models by the TBS dimension—applied on a multi product process chain in the railway industry. Accepted for 21st CIRP LCE. Trondheim, June 2014

Rohrmus D, Holst J-C, Mose C, Müller K, Schedlbauer M, Raschke U, Weinert N (2011) Sustainable manufacturing at Siemens AG. In: Seliger G, Khraisheh MMK, Jawahir IS (eds) Advances in sustainable manufacturing. Springer, Heidelberg, Dordrecht, London, New York, pp 79–84

Weinert N, Fischer J, Posselt G, Herrmann C (2013) Lean and green framework for energy efficiency improvements in manufacturing. In: Seliger G (ed) Proceedings of the 11th global conference on sustainable manufacturing. Innovative Solutions, pp 512–516

Chapter 6
Cyber-physical Approach for Integrated Energy and Maintenance Management

Benjamin Neef, Christopher Schulze, Gerrit Posselt, Christoph Herrmann and Sebastian Thiede

Abstract Because of nowadays complex and highly automated industrial production lines, every stoppage involves the danger of a massive economic harm. That's why companies use already various production, quality and maintenance methods to reduce—or at least to handle—unforeseen stoppages. This paper presents a novel approach to improve the reliability of production fields by supporting predictive maintenance under the combination of systems from energy and maintenance management. Wireless sensor networks and mobile devices are integrated into a cyber-physical system to gain real-time transparency of energy demands within production environments. Being aware of challenges introducing cyber-physical systems into the brownfield, the proposed solution considers needs of data standardisations, IT security, staff participation, big data handling, long-term technical risk and cost-benefit estimations. The developed methods are considered by user-oriented design principles to deliver role-specific information. Therefore, the derivation of these informational requirements is based on production unique job activities. Allocating time and component-based energy demands whilst taking machine and environmental conditions into account enables a basis of comparison and a continuous improvement process of energy efficiency and maintenance. These demands are fulfilled by the methods of a continuous energy value stream mapping, an energy efficiency tracker and an integrating energy and maintenance monitoring. This proposed approach is based on the ESIMA project funded by the German Federal Ministry of Education and Research. The project aims for "Optimised resource efficiency in production through energy autarkic sensors and interaction with mobile users".

B. Neef (✉) · G. Posselt · C. Herrmann · S. Thiede
Institute of Machine Tools and Production Technology,
Technical University Braunschweig, Braunschweig, Germany
e-mail: b.neef@tu-braunschweig.de

C. Schulze (✉)
Daimler AG Technology Management, Motor Factory, Mannheim, Germany
e-mail: christopher.schulze@daimler.com

S. Thiede and C. Herrmann (eds.), *Eco-Factories of the Future*, Sustainable Production, Life Cycle Engineering and Management,
https://doi.org/10.1007/978-3-319-93730-4_6

6.1 Introduction

Apart from the enhancement of conventional production-related objectives like cycle time and quality, the improvement of the efficient use of energy is becoming one of the most relevant industrial developments. This is mainly driven by the European ambitions to reduce energy demand as well as carbon dioxide emissions by 20% compared to 1990 until 2020 (European Commission 2018). With a share of around 30% of Germany's energy demand, the industry (Umweltbundesamt 2018) is forced to contribute by improving their own energy utilisation. In addition to legal requirements the rising energy demand, rising energy prices and sophisticated environmental requirements of the consumer enforce the enterprises to develop.

One leading movement in factories is the increasing digitalisation and interconnection between machines, which allow a higher transparency, performance and flexibility of production processes. This progress is initiated as a part of the high-tech strategy proclaimed by the German federal government in 2011 (Bundesministerium für Bildung und Forschung 2018). The vision of the fourth industrial "revolution" is essentially impelled by the technological concepts of cyber-physical systems and the Internet of things (Hermann et al. 2015). The consolidation of both is supposed to enable real-time communication between human–machine and machine–machine.

Expecting to benefit from the assumed potentials, new fields of research are coined, for example in order to identify synergies between energy management and predictive maintenance. The subsequent approach combines the aspects of machine condition and energy monitoring under consideration of a minimally invasive integration into the brownfield.

Thereby, this paper describes efforts to design and develop software-assisted methods for the analysis and visualisation of energy and machine data. A particular focus is being placed on an approach for predictive maintenance and also in order to provide an energy monitoring and management system. Concrete ambitions are for example:

- To support a continuous improvement process, a certain concept for evaluating measures of improving energy efficiency and to support maintenance activities was developed.
- By visualising and overlaying the machine data and the energy demand, a higher transparency is provided concerning the correlation between the operating modes and the energy demand.
- By energy value stream mapping, productive and unproductive rates of produced components can be revealed.

However, due to their high variety of action, the sense of responsibility and the individual reaction ability, humans in the production environment will continue to be the master control instance of the system within the cyber-physical casual network (Bauernhansl et al. 2014a, b). As shown in Fig. 6.1, the human (employee) is centred between the physical and the virtual components. System behaviour of the physical components becomes more transparent and replicable for the employee by the compound and information exchange of software elements within the cyber (virtual and

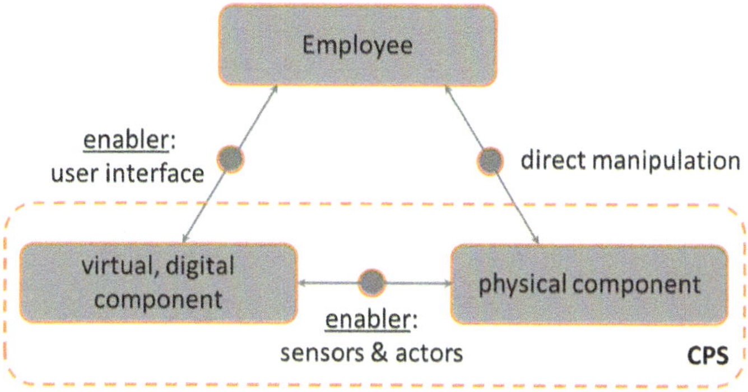

Fig. 6.1 Cyber-physical system casual network (Gorecky and Loskyll 2014)

digital components) and physical components. Gorecky et al. describe that either direct interaction takes place between the employee and the physical component or indirect manipulation of the physical components takes place across the virtual, digital component (Gorecky and Loskyll 2014).

Thus, the software-based visualisation and interaction of the developed methods were compiled with regard to human centred designed principles (DIN EN ISO 9241-210 2011). Therefore, the users of these approaches were placed at the centre of the development process to reach a high usability (Chamberlain et al. 2006). Production workers, energy managers, maintainers, team leaders, operating engineers and managers have been identified as beneficiaries of the above-mentioned solutions, since they will be enabled to assess energy efficiency measurements and to facilitate condition-based maintenance strategies.

6.2 Challenges by Introducing CPS into Brownfield

Apart from these potentials, enterprises are facing various obstacles, regarding Fig. 6.2, when introducing cyber-physical systems into brownfield environments. The following listing gives an overview of challenges that are arising while creating an increasing interoperability of different systems by merging production and information technologies (DFKI 2014):

- **Lack of standardisation of data and interfaces**
  In addition to the known difficulties, such as underdeveloped standards for data interfaces and communication protocols, a practical realisation rises to following challenges: on the one hand, for the implementation of interfaces in consisting systems, experts with knowledge of often outdated programming languages are needed. On the other hand, bureaucratic obstacles have to be overcome in order to gain access to a running and established system. The inability in mastering

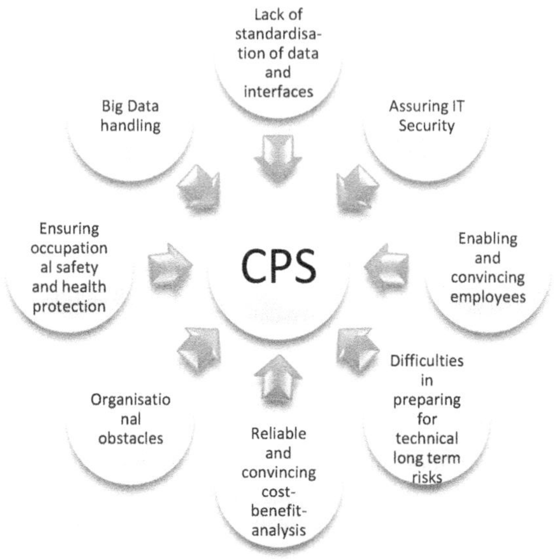

Fig. 6.2 Challenges for cyber-physical systems

the system complexity as well as the common stable opinion of "never change a running system" substantially impedes progress.

- **Assuring IT security**
  The development of cyber-security and data protection is of particular importance for the industry (Ziesemer 2015). Especially, frequently arising issues concerning topics like the Internet of things and cloud computing cause distrust towards IT applications.
  Due to the increasing interconnectedness of systems, the production and business IT have to face higher requirements for security technologies, such as anti-virus protection, firewalls, virtual private networks and user authentication (Fallenbeck and Eckert 2014). According to Fallenbeck and Eckert, security gaps pose tremendous risks even potentially for the physical world, e.g. real physical harm can originate from hacked production equipment. High standards and real-time conditions of industrial systems need to be fulfilled here. In order to obviate or at least reduce these risks, it is necessary to implement innovative security solutions, which is particularly difficult in brownfields, as they were once designed for a long-term use (>20 years) and in no case dimensioned for the safety aspects of today (Ganschar et al. 2013). Especially, the implementation of cryptographically protected network interfaces and a comprehensive authorisation management is affected, because user authentication, encryption and decryption of data need to be done in a minimum of time (Fallenbeck and Eckert 2014).
- **Enabling and convincing employees**
  On the one hand, introducing new technologies causes insecurity due to a lack of knowledge and overextension of some employees. On the other hand, it leads to

distrust and resistance of employees, who consider their knowledge of their personal property and assume their jobs being endangered by an increasing automation. While a lack of knowledge can be addressed by qualification and training measures, a lack of acceptance is very difficult to overcome. The technological development from an exclusive data preparation to a self-organising and decision-making system is often construed as an ethically critical development in which the employee, the former controller of the system, becomes a system-controlled production good. In order to avoid or at least slow down this process, some employees eschew to feed in their knowledge into the system.

- **Difficulties in preparing for technical long-term risks**
  Due to insufficient possibilities of technical testing, a risk estimation of industry 4.0 technologies is yet rather challenging, especially in the long term. Therefore, the constant availability of infrastructure and system components (e.g. WLAN by the use of new wireless communication technologies; keyword: denial of service) as well as the compatibility with existing highly diverse components and systems within brownfields cannot be assured adequately.
- **Reliable and convincing cost-benefit analysis**
  At present, cost-benefit analysis of industry 4.0 technologies either cannot be made adequately due to a lack of experience or cause caution because of an insufficient calculated profitability. In particular, concerning energy costs, high investments often do not simultaneously imply high savings. Especially for companies that are mainly driven by short-term goals, the payback period of energy-saving models is too long.
- **Organisational obstacles**
  From an organizational point of view, a lack of responsibilities and inadequate management support are the main barriers for the adoption of industry 4.0. The cross-department cooperation is indispensable for a successful implementation of complex systems and applications with widespread information demands.
- **Ensuring occupational safety and health protection**
  At times, implementing new technologies could affect occupational safety and health protection notelessly. For instance, the long-term effects of using data eyeglasses are yet not investigated adequately. Conceivably, they could distract the employee or even cause fear of surveillance, which could possibly endanger their safety and health.
- **Big Data handling**
  Filtering and aggregation of a various big data volume for a comprehensive, close to real-time and reliable use (Tole 2013) are a relevant challenge of industry 4.0. Nevertheless, large companies already reached a high level of data acquisition in comparison to small and medium-sized enterprises, which has been proven by established systems like monitoring and management of power screwdriver data, pick to light and laser-based commissioning.

6.3 Concept Development

Based on the rapid developments in Information and Communication Technology (ICT), new opportunities open up, as well in manufacturing environments. Cyber-physical systems (CPSs) offer direct interaction possibilities between virtual software-based systems ("cyber-world") and physical objects ("physical world"), e.g. manufacturing equipment. Interaction between different components of a CPS is provided due to wired or wireless communication networks. In conjunction with an energy management system, assistance to achieve energy objectives and policies by interrelated or interacting elements is provided. By monitoring of energy demands and key performance indicators (KPIs), organisations are able to save resources and to obtain a financial benefit (Posselt et al. 2014). By incorporating process and PLC data an integrated maintenance management system supports advanced scheduling activities and reduces machine failures.

Figure 6.3 shows a framework of cyber-physical (production) systems with the four subsystems (I–IV), their single elements and the necessary interfaces (a–h). Information transfer between the physical and the cyber-world is done within a closed control loop including different means for data acquisition and storage, appropriate models and decision support or even automated control schemes. Data acquisition on physical level could be realised by energy self-sufficient sensor nodes placed beside or in a machine's electric cabinet and additionally in the environment to monitor production conditions. To reduce costs and expenses for wiring, data acquisition on field level could take place with energy-autonomous sensor nodes and wireless data transfer (Neef et al. 2017).

The human stays in the centre of attention, as the operational instance in the real world and the beneficiary of the provided methods and tools provided by the

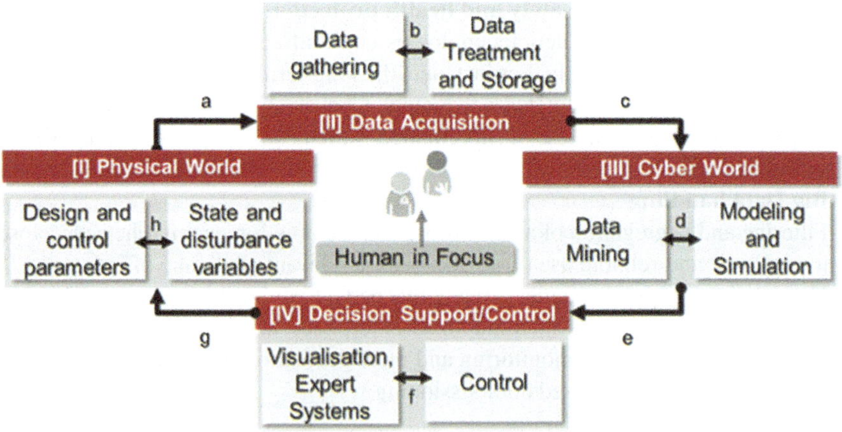

Fig. 6.3 Functional diagram of a physical production system and mapping into a virtual environment (Thiede 2018)

cyber-layer. Hence, within a factory environment, different user roles have to be distinguished since they incorporate different tasks, responsibilities and qualifications. Thus, in context of the envisioned integrated energy and maintenance management, they need to be addressed through different means. Those roles are derived and described in the following section.

6.4 User Roles and Derivation of Information Requirements

Production-related roles can be found in almost every large manufacturing enterprise. In general, these user roles differ in the amount of strategic and operative tasks they have to fulfil. While management positions also have strategic tasks, executive roles mostly fulfil operative tasks. Staff responsibilities and the authority to make strategic decisions are the main characteristics for management positions. The energy manager as a staff position gives support for strategic decisions along all hierarchy levels while also executing management decisions concerning energy efficiency.

The higher a role is classified within the hierarchy, the higher is the amount of his strategic tasks and the more aggregated the information need to be for him. Vice versa, the more operative the daily work, the higher is the employee's demand for detailed information concerning his specific tasks.

Therefore, the information demand of a specific role depends on their positions and activities. Some identified professional activities of these user roles and the derived informational requirements of an integrated energy monitoring and maintenance approach are shown in the following tables. Table 6.1 displays some main activities and informational requirements of management positions. In order to plan operational and strategic measures under consideration of their personal and financial resources, these leadership members need aggregate information and figures on the current situation and expected developments with regard to energy and maintenance issues. Meanwhile, the CEO and managers focus on corporate objectives; team leaders and foreman are measured on achievements of specific production domains. All of these management members can delegate tasks to their subordinated employees, e.g. operators.

Table 6.2 have their high operative activities and mostly immediate influence on production in common. The operating engineer has a more future oriented planning focus in comparison to the maintenance employee and the machine operator, which both have a more day-to-day driven business.

Table 6.1 Management positions, activities and informational requirements

	Activities	Informational requirements
CEO	• Defining corporate goals for energy and maintenance issues	• Corporate energy demand and CO_2 emissions
Manager	• Planning and decision-making of strategic targets in consideration of corporate goals • Providing production forecasts • Initiating and executing change processes	• Aggregated key figures (energy demand development, maintenance activities) of specific department • Graphical preparation of trends
Team leader	• Fulfilling operative targets • Developing, coordinating and implementing of maintenance concepts • Qualify and promote his employees for maintenance and energy issues	• Aggregated key figures • Energy value stream of assembly lines • Environmental conditions (illumination, temperature, etc.)
Foreman	• Coordinate repair orders and employees • Define measurements of manufacturing, assembly and maintenance plans • Adhere deadlines of maintenance • Problem analysis and forward of recommendations for process improvements • Evaluate maintenance measures	• Assembly line energy demand Energy value stream of assembly line • Individual KPI for energy efficiency and maintenance-based failure safety

Table 6.3 exemplifies the abundance of the activities of an energy manager. As an expert for energy efficiency, his informational demand focuses on energy issues.

6.5 Introducing Methods and Tools

Figure 6.4 depicts the identified user roles with the derived specific informational requirements. Depending on the hierarchical position of a user role, varying demands concerning the granularity of information must be considered. Consequently, user role depending aggregation of data and information is requested and provided by customisable key figures and meaningful charts. The user-specific informational requirements are drawn on methods and tools that will be presented in the further sections. An essential enabler to apply the derived methods and tools is the aspiration to allocate reproducible time and component-based energy demands (compressed air and electricity) during processing while taking machine and environmental conditions (e.g. temperature, humidity, luminous flux, carbon dioxide and airflow) into

Table 6.2 Executive positions, activities and informational requirements

	Activities	Informational requirements
Operating engineer	• Achieve manufacturing and maintenance targets under strategic guidelines • Analyse and identify effects of assembly lines and possible weaknesses • Realise optimisation measures • Create service specification for utilities and machines	• Energy value stream of assembly lines and machines • State surveys of machines and assembly lines • Flexible demand measurements and documentation
Maintenance employee	• Ensure machine uptime • Maintenance (shutdown, inspection, servicing, repair, optimisation, recommissioning) • Supervise condition monitoring and maintenance management system • Analyse and remedy error and weak points • Optimise utilities and machines Document and track performed measures	• Graphics of assembly lines and machines structures • Customisable key figures and trends • Measurement data overview • Sensor infrastructures • Machine state-based energy demand (also after maintenance activities) • Machine control data
Machine operator	• Analyse work pieces and machine faults • Remedy by low complexity and ordering maintenance by serious faults/failures/errors • Partially preventive maintenance and repairing • Documenting of faults and failures	• Machine and energy states

account. The presentation of method and tool-based computational outcomes can be provided on mobile devices, e.g. a tablet pc or smart phone.

6.5.1 Allocation of Specific Energy Demand as an Enabler

Energy demand of a machine tool is highly dynamic and depends on the interaction of different machine components which results in distinctive machine states. Assignment of energy demand to components or an allocation to the originator is highly dependent on clear and meaningful data sets, particularly to assign energy or auxiliary demand to products. The United States Environmental Protection Agency (US EPA) firstly introduced the allocation of energy use to products as an extension to

Table 6.3 Energy manager, activities and informational requirements

Activities	Informational requirements
• Achieve corporate energy objectives by energy efficiency optimisation and CO_2 reduction measures • Supervise energy monitoring and management system • Review current state energy situation • Identify and analyse energetic wastage and derive optimisation measurements • Consult management for planning and realisation of energy objectives • Develop and implement an optimal resource usage • Accomplish economic feasibility studies of machines and processes • Evaluate possible invests in energy optimisation • Benchmark machines and processes	• Key figures and trends • State-based demand of machines and assembly lines • Energy demand and CO_2 emissions of departments and whole company • Track of energy efficiency measures and energy optimisation process

The informational requirements of these user roles need to be fulfilled by the methods and tools presented in the following sections

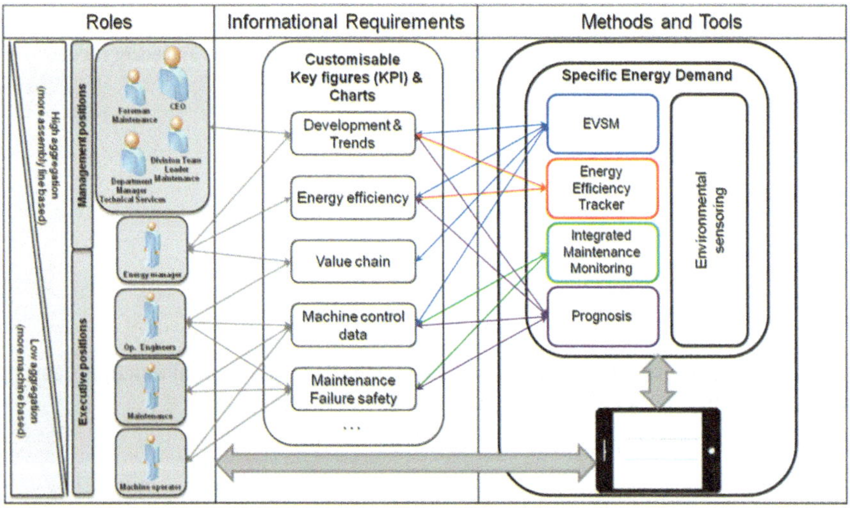

Fig. 6.4 Linkage of user roles, informational requirements and methods and tools to achieve requested requirements

the existing value stream mapping methodology. By adding energy aspects to value stream mapping, energy costs can be reduced and the productivity can be improved (US EPA 2011).

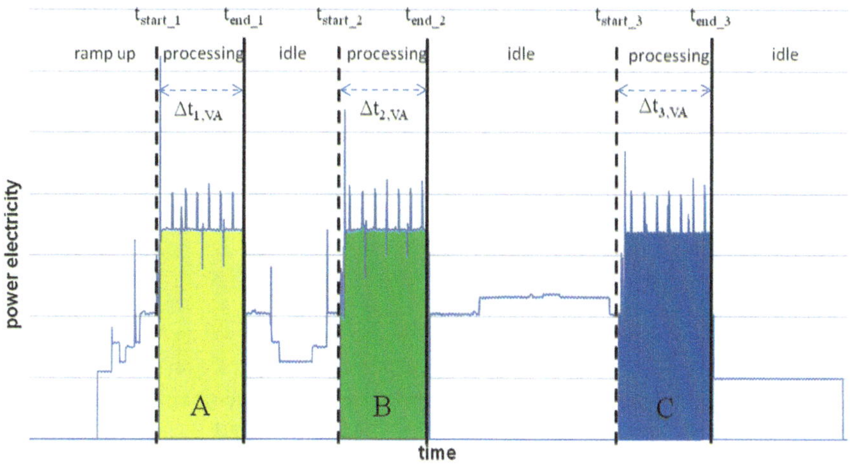

Fig. 6.5 Schematic allocation of electricity demand for identical products

As stated above, production processes in general require different energy carriers and auxiliary supplies to create value. The demand and type of media to operate the production process vary over time. The production cycle can be principally distinguished into value-adding and non-value-adding times (Erlach and Westkämper 2009). If the supply of the production process is not disconnected, there is typically a certain amount of energy and periphery media needed to keep the process in ready for operation state. This time is called non-value-adding time that can take place in different machine states, e.g. during standby (e.g. machine idling while waiting for parts), maintenance or failure.

The use of real-time machine data combined with metering data of energy and auxiliary flows to allocate energy and auxiliary demands to discrete products is the key in deriving meaningful performance figures and characterise industrial processes. The unique assignment of energy use to certain products during the value-adding time makes the process itself comparable. Figure 6.1 Cyber-physical system casual network (Gorecky and Loskyll 2014)

Figure 6.5 depicts an exemplary load profile and indicates graphically how the allocation of the value-adding electricity demand can be performed. The dashed vertical line represents the starting time of the value-adding process (processing). The solid line represents the end time of the value-adding process (and start time of the non-value-adding process time during machine's idle mode). Both events can be read from machine's programmable logical controller (PLC) or production planning systems. The energy (e.g. electricity and compressed air) and auxiliary (e.g. lubrication, water) demands for identical products (e.g. area A, B and C) are reproducible within specified tolerances. Therefore, the part-specific measured demands are suitable for energy efficiency monitoring and tracking of energy efficiency measures.

To reduce internal costs of energy for the manufacturing industry, the implementation and tracking of energy efficiency measures as a part of a continuous improvement

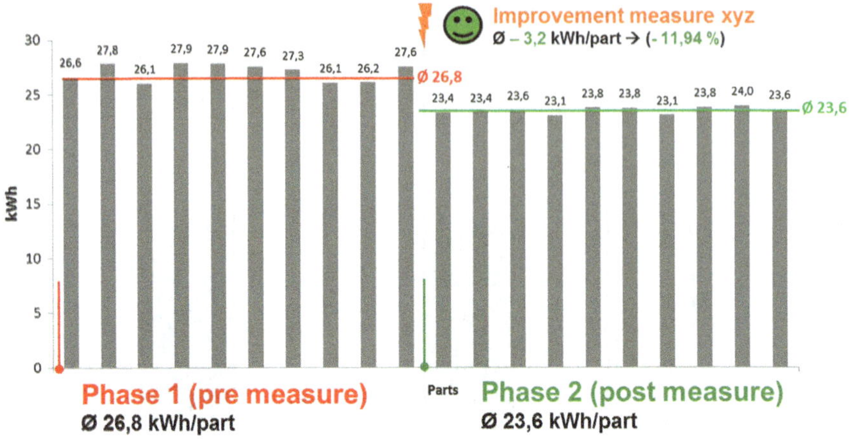

Fig. 6.6 Monitoring for an energy efficiency measurement with part-specific energy demand

process are an indispensable part. To establish a continuous improvement process, the possibility to evaluate implemented energy efficiency measures is a fundamental condition (May et al. 2015) Adequate reproducible key performance indicators should be periodically reviewed to conform to the requirements of a continuous improvement process, e.g. within the framework of ISO 50 001 energy management (DIN EN ISO 50001 2011–12).

To support this systematic approach, different interactions and evaluation possibilities drawn of the collaborative data pool were implemented. According to the causative principle, the distinction between different energy carriers and the examination by different machine states is an important aspect. To achieve intended energy efficiency goals, the user can set milestones with information concerning time, type (e.g. compressed air or electricity), reference quantity and tracing of the applied energy efficiency measurement. By the use of steadily monitored mean values, the user can spot the achievement of target values and acting up to the ISO 50 001 procedure "plan, do, check, act" will be simplified.

Figure 6.6 shows exemplarily the layout of a monitoring view for energy efficiency measures. The view includes the mean value of part-specific value-adding energy demand. The essential outcome is indicated in the row "optimising measure xyz" and shows the decrease in value-adding part-specific energy demand.

6.5.2 Dynamic Energy Value Stream Mapping

Energy value stream mapping (EVSM) evolved out of value stream mapping methods developed for lean manufacturing purposes. The extension by the energy dimension allows the designing of time and energy-efficient production systems simultaneously.

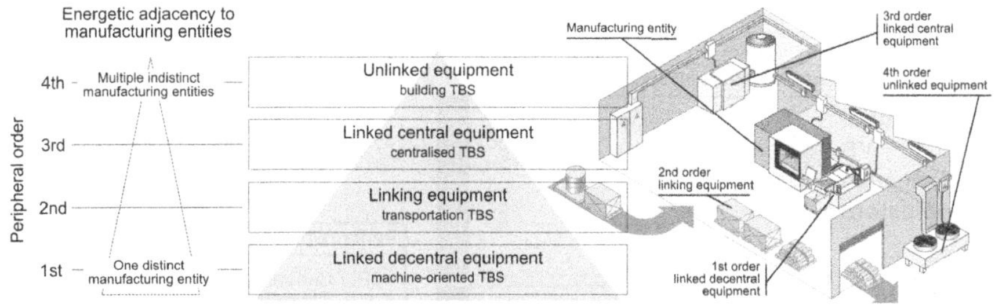

Fig. 6.7 Energetic adjacency of TBS to manufacturing entities (Posselt 2015)

A wide range of possible approaches can be found in the current state of research (US EPA 2011; Erlach and Westkämper 2009), mainly addressing the same basic set of key performance indicators extended by their specific performance indicators to ensure a more holistic perspective on energy flow transparency. Bogdanski et al. introduced an extended energy value stream approach that includes the technical building service (TBS) dimension (Posselt et al. 2014; Bogdanski et al. 2013). As shown in Fig. 6.7, Posselt further distinguishes the energetic adjacency of TBS providing equipment in relation to the value-adding manufacturing entity (Posselt 2015).

The comprehensive assessment regards energy shares demanded by direct value-adding core processes of production machines up to unlinked equipment with lowest possible adjacency to the core process as, for example, floor heating systems. The used energy for the value-adding process time of the core process E_{VA} can be determined as a function of the arithmetically averaged power during value-adding $\bar{P}_{VA}$ and the sum of the work piece-specific value-adding time $\Delta t_{w,VA}$:

$$E_{VA} = \bar{P}_{VA} \cdot \sum_{w=1}^{k} \Delta t_{w,VA} \tag{6.1}$$

Further information about calculation for non-value energy and consumers on higher peripheral order can be found in (Posselt et al. 2014). Figure 6.8 depicts the considered elements up to the fourth order in an energy value stream box (Posselt et al. 2014). With this holistic assessment perspective, a true cause and effect analysis can be achieved through creating energy value stream maps in a reoccurring, comparative manner.

So far, the methodology of EVSM was widely applied in a pen and paper-based manner, representing only a temporal snapshot of a current condition which comes with an evident pair of disadvantages as shown in Table 6.4.

The concept introduced in Fig. 6.9 now offers the possibility to implement the EVSM methodology within a dynamic data stream that is based on a collaborative data pool within the cyber-layer. The new cyber-physical approach allows having a total process chain perspective and hence preventing the phenomenon of problem shifting caused by local improvement activities.

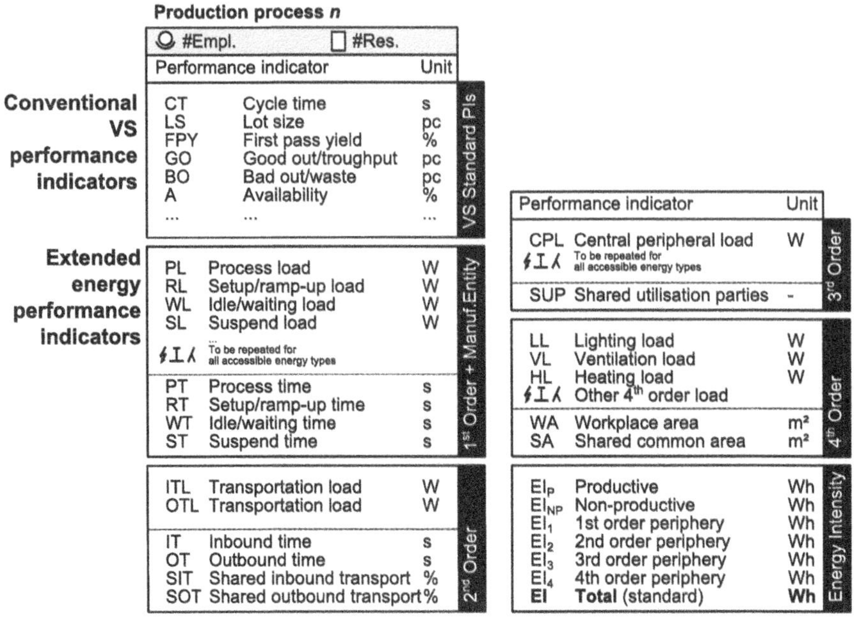

Fig. 6.8 Energy value stream box (May et al. 2015)

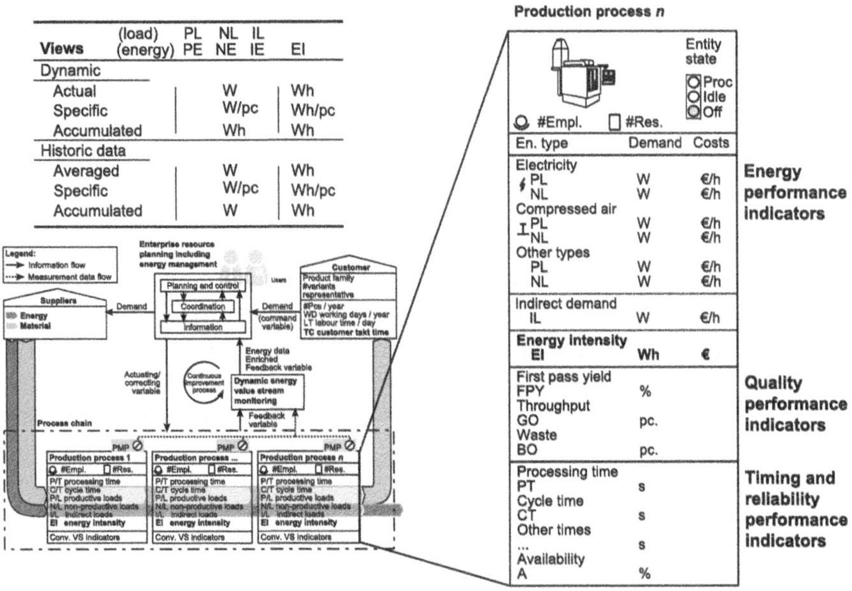

Fig. 6.9 Visualization of energy and conventional value stream KPIs (Posselt 2015)

Table 6.4 Disadvantages of the static pen and paper-based approach and new specifications of the dynamic approach eliminating these drawbacks

Disadvantages of static pen and paper EVS mapping	Specifications of new dynamic EVS monitoring
No differentiation of productive and non-productive time shares	Application of intelligent PMPs (IPMP) to automatically detect entity states
Before and after acquisition of EVS maps requires a repetitive, time-intensive measurement	Low effort to create EVS maps from dynamic data streams
Improvements in the system need yet another temporary measurement to evaluate possible benefits	Immediate indication of differences between two EVS maps, in an accumulated and process quantity view
Interdependency effects between entities in the linked process chain are not intuitively identifiable	Intuitive visualisation to identify interdependencies
Different aggregation periods rely on extrapolation of temporary short-term measurements	Reliable, full-term, parallel monitoring of actual power demands and calculatory integration (tool supported)
Support of a continuous improvement process is cumbersome	Easy to use and quick to apply approach to support continuous improvement actions and reporting

The dynamic energy value stream mapping procedure has to follow a standard set of calculation rules for the desired performance indicators in order to obtain reproducibility. All entities monitored have to be describable with a consistent set of entity states in order to allocate productive and non-productive energy states (energy effectiveness). To estimate the energetic and economic effect of measures on longer periods of time, simple extrapolation functions must be incorporated. To consider peripheral entities of different orders, the metering strategy has to be extended accordingly. In order to ensure quick implementation and later extension, common industrial interfaces of energy flow sensor shall be considered (e.g. standard interfaces for three-phase electric power, analogue inputs for non-electrical energy flows). To support the intuitive understanding of the KPI evaluation, a dynamic visualisation with the already present layout and design of the material-oriented value stream must be realised.

The energy performance indicators in the category below give detailed information about any of the utilised energy types (external and internal ones). The values and units displayed vary, depending on the selected view. The user can decide on a dynamic view showing actual values (process quantities), specific process quantities in a reference unit of one piece of processed goods or the accumulated values since the beginning of the accumulation period (e.g. start of a test series). The user can, at any time, save the set of specific and accumulated views to a database for later comparison. The accumulated productive energy PE is calculated for each energy type as $PE = PL \cdot PT \cdot GO = [\mathrm{Wh}]$.

The accumulated unproductive energy NE is calculated for each energy type as

$$NE = NL \cdot (CT - PT - OT) \cdot (GO + BO) = [\text{Wh}],$$

with OT being the off time share within the CT of a resource. The indirect energy IE for the case of the fourth peripheral order is calculated as

$$IE = IL \cdot CT \cdot GO = [\text{Wh}].$$

Correspondingly, the key energy performance indicator of a process EI is calculated as the sum of productive- and unproductive- and also indirect energy shares:

$$EI = PE + NE + IE[\text{Wh}],$$

and the specific key performance indicator as

$$EI_{\text{spec}} = (PE + NE + IE)/GO[\text{Wh}].$$

Saved data is regarded as historic data. It serves as a basis for a series of EVS map comparisons to accompany the start of production of new products, or for the implementation of energetic or conventional improvement measures. The saved historic views can be printed or directly compared by difference indication to allow the identification of bottleneck processes, energetic drivers and different types of lean and energetic wastes (e.g. idle times, high idle loads, uneven levelling of process performances, overproduction, unproductive energy demands and ineffective utilisation of energy types).

For the utilisation of the dynamic EVS monitoring, a metering strategy is to be applied by means as described by Posselt (Seera et al. 2014). The focus can be narrowed down to the entities of the desired process chain, but to be able to incorporate the indirect energy demands (I/L), peripheral entities are recommended to be included by making sure to define the source–sink relationships as well as the energetic adjacencies of each manufacturing entity.

Hence, the capability to interact with real EVSM live data is given. Computation results and consequential recommendations for action are displayed target-oriented as a control strategy or as a recommendation for a specific action within e.g. a mobile application. As a result, energy value stream mapping becomes easier when using a consolidated ICT system as introduced. The presented ICT system covers the relevant data path from shop floor level to the specific user. Three stages of data transfer can be distinguished:

- **Data acquisition on field level**:
  To meet the requirements of the presented methodology, data acquisition on field level is necessary. Power measurement of the actual power of the core process and the attached peripheral units with an update frequency of one second and as well event based process information read from a machine execution system

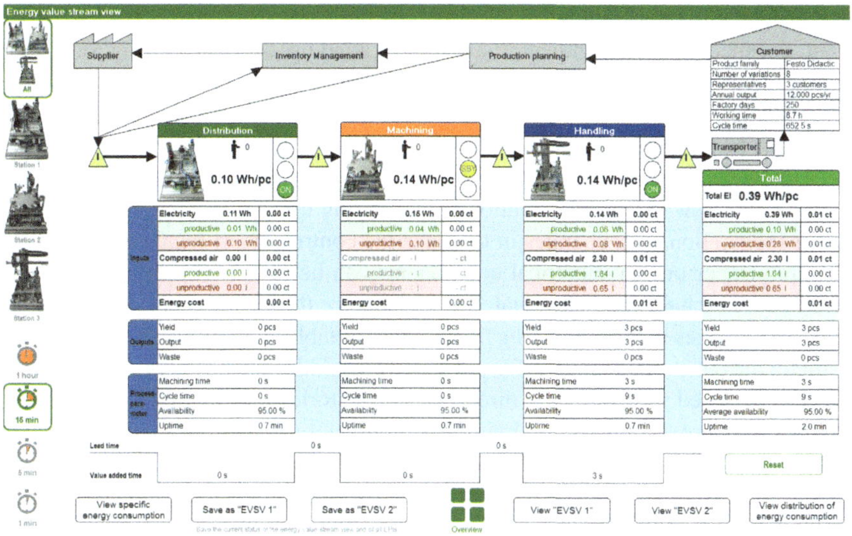

Fig. 6.10 Dynamic energy value stream monitoring applied on a *Festo Didactic* modular production system for educational purposes (Posselt 2015)

(MES) or from the machine's programmable logical controller (PLC) represent the collaborative data basis for computation.

- **Preprocessing and computational analysis within the cyber-layer**: The cyber-layer represents the unit to synchronize and perform syntactic analysis of raw process and measurement data. Depending on the available prepared data from field level, several methodologies—besides EVSM—could be provided within the cyber-layer to satisfy the informational requirements of the user. Calculation algorithms, e.g. Eq. 6.1, are implemented within a backend unit.
- **(Mobile) presentation within value stream boxes**: The mobile presentation represents the interface to the user, e.g. the mobile maintainer. Target-oriented indicators to evaluate manufacturing processes (calculated within the cyber-layer) are graphically prepared and displayed. Figure 6.10 depicts schematically a structure outline to display EVSM specific performance indicators.

The implementation of the methodology and dynamic performance indicator calculation is realised on the PLC of the energy transparent machine hardware concept, but with a higher performing processor for the operating system of the IPC operating the PLC. The main view of the developed frontend is shown in Fig. 6.10, depicting the typical value stream map and the new energy-aware performance indicators.

As soon as the application is started, the status indication of the three stations comes up. Upon the activation and initialisation of the stations, the idle signal indicates ready for production. The idle load is indicated, and for the aggregated indicator view, the non-productive energy share integrates the load for each station, differentiated into its supplied energy types. As soon as the first work pieces are processed and throughput is generated, the resulting OK parts are accounted as yield. At this point, the specific energy indicator set can be called up, as shown in Fig. 6.10. The energy intensity per part is accounted in the header of each station. At the bottom, the productive and idle time shares are indicated.

On this basis, the users can make test runs with a representative number of work pieces to create a benchmark and save it as a static energy value stream map in the history data. Now technical alterations by changing the compressed air pressure at the air preparation unit, changes in the machine control, smart local controls of the vacuum gripper or new electrical components can be tested. The influence on the whole process chain can be evaluated at an instance through higher or lower actual loads and changes in the processing times. Most suitable configurations can be tested again for the benchmark number for yield, and a secondary static energy value stream map can be saved and compared directly to the historic benchmark.

It has been proven in the experimental demonstration that the dynamic and instantaneous feedback of energy performance indicators is highly important for a continuous improvement process. Furthermore, the effect of a single parameter on the whole picture (in this case the process chain, in the real case the whole factory) is important to understand. Without instant informational feedback, a cause and effect relationship can hardly be generated by the involved persons. Especially, when it comes to more complex setups, it is recommendable to also include the direct energy in the dynamic monitoring system, as it helps to integrate facility and production domains with one tool (Posselt 2015).

6.5.3 Integrated Energy and Maintenance Monitoring

The German Institute for Standardisation defines maintenance as the "combination of all technical, administrative and managerial actions during the life cycle of an item intended to retain it in, or restore it to, a state in which it can perform the required function" (DIN EN 13306 2010–12).

From an economic point of view, a fundamental goal in maintenance is to keep the reliability of a facility at a high level while simultaneously optimising the overall costs consisting of maintenance and downtime costs as depicted in Fig. 6.11 (Herrmann 2010). The left side represents breakdown maintenance strategies. Breakdown maintenance ensures the functional integrity of production machines by intervention in case of breakdown. In contrast, preventive maintenance follows the strategy of scheduled preventive replacement of abrasion-prone parts before a breakdown occurs (Reichel et al. 2009). This approach is located on the right section of Fig. 6.11 and is characterised by minimal downtime costs.

The optimal maintenance strategy tagged with the black dashed line is based on the knowledge when a maintenance action should be performed and represents the most economically reasonable solution. To operate a maintenance workforce at this optimal point, either real-time information about the condition of a component achievable by ad hoc monitoring is required or an integrated system can be set up.

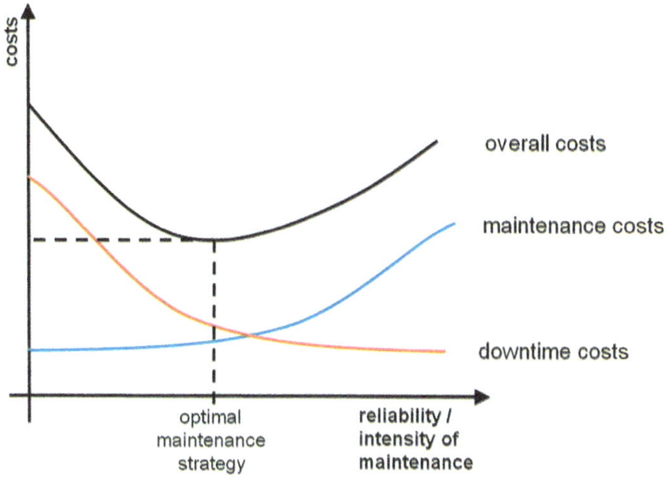

Fig. 6.11 Optimal maintenance strategy (US EPA 2011)

By setting up an integrated system, a collaborative data pool to obtain necessary information can be used. The information provided by this data pool forms the basis for several methodologies. In this manner, the necessary data is already existent within the cyber-layer and is part of the closed-loop system introduced in Fig. 6.2. Additional measurement technology or equipment is no longer required. Nowadays, computational performance of an embedded system is sufficient to use data of simple measurement technology, e.g. electricity measurement to gain further information about the condition of a component or a machine.

In this context, the correlation between current signature and wear out of machinery and components is scientifically proven by several authors. Seera et al. detected three different common induction motor faults by using real current signals (Seera et al. 2014). Anwar K. Sheikh et al. experimentally state the correlation between drill wear out and increased power demand of a milling machine (Sheikh et al. 2005). Sami Kara et al. demonstrated the correlation between operating time of electric drives (for the case of washing machines) and several state variables. He could prove that the rotational motor speed, the power demand, vibrations and temperature of the motor were reasonably related to the age of the motor (Kara et al. 2005).

Assume that the specific demands of electricity, compressed air and lubrication per produced part follow a normal distribution $N(\mu, \sigma)$ with mean μ and standard derivation σ over the time as shown in Fig. 6.12. The mean will be calculated by a finite set of time independent measurements of energy or auxiliary demand. As a result, a discrete sequence of mean values will be achieved. Occurring systematic measurement errors will be equalised. Derivations of this mean values can be detected automatically. The abnormal change of the mean energy demand per part (e.g. as a result of machine wear out over time) is indicated by the dashed vertical red line.

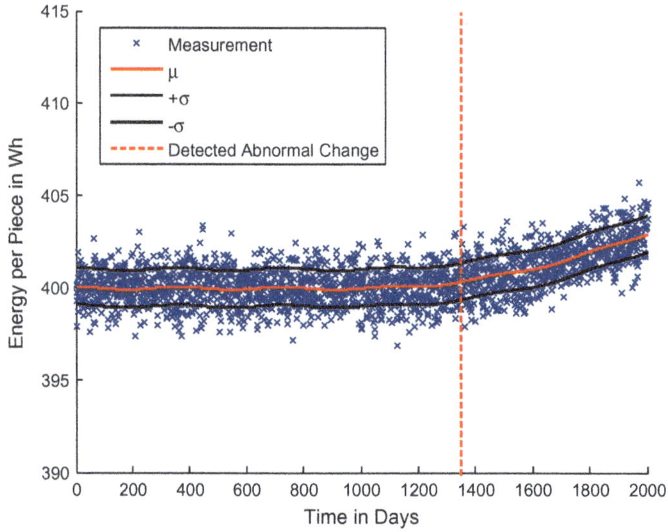

Fig. 6.12 Change in energy demand per part over time and detected abnormal threshold

6.5.4 Prognosis

A reliable estimation of a components condition is naturally a necessary but also not sufficient condition to improve maintenance activities. Additionally, the prediction of possible future states of a component and its overlying system is crucial. Studies indicate that for the estimation the actual operating time of a component, different state variables (e.g. vibrations, temperature, voltage or electrical power demand) in combination with methods like regression analysis or artificial neural networks can be used (Kara et al. 2005). Based on these values, also an estimation of the remaining lifetime and, thus, potential maintenance demand can be given. However, this mainly relates to the exchange of components as maintenance strategy and also neglects the stochastic nature of the problem. Simulation, and more specifically Monte Carlo simulation, is an appropriate approach here. Based on values like mean time between failure (MTBF) that can be derived from the above-mentioned data analysis, failure rates and failure probabilities can be calculated and used for the simulation. To achieve the necessary statistical robustness, a sufficient quantity of simulation runs needs to be conducted which results in a distribution of result values. Figure 6.13 shows the example of an industrial robot with four different drives. In this case, the total cost of ownership (TCO, includes maintenance, failure and energy costs) was used to compare the effect of different maintenance strategies—breakdown, periodic and condition-based maintenance with varying intervals were analysed (Herrmann 2010). For decision support, the average value over all simulation runs (here 385 runs per scenario) but also the spread of the values as measure for the risk of a strategy is of importance. It gets quite clear that maintenance strategies significantly differ in

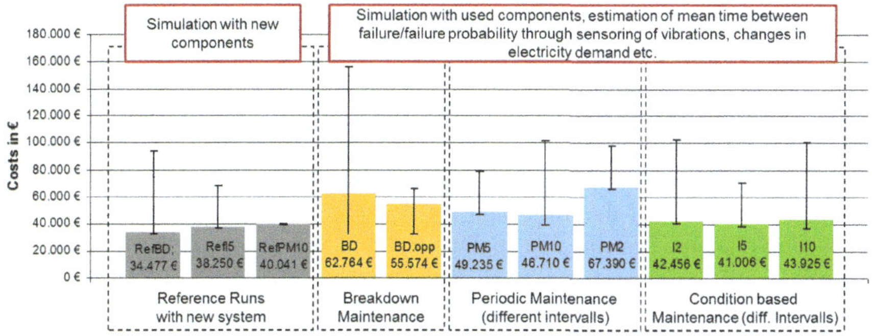

Fig. 6.13 Exemplary results of Monte Carlo simulation of different maintenance strategies (case study for an industrial robot) (Herrmann 2010)

terms of their impact on component behaviour and, thus, related TCO. The example also underlines the necessity to consider the specific circumstances of the situation; for example, a new system should be maintained different compared to an operating system where all components show varying conditions.

Data-based analysis and prognosis functionalities can give distinctive decision support for maintenance. On the one hand, the example underlines the potentials, but on the other hand, the inherent complexity and simulation efforts are certainly a barrier for industrial application on a continuous base. However, being clearly related to current developments in context of cyber-physical systems, increasing IT performance and intuitive applications will allow to conduct those simulations in daily work and even decentralized on smart devices.

6.6 Critical Review and Outlook

The described solution gives reason to expect synergies for predictive maintenance and energy management, as the integrated energy and condition monitoring system achieve a high transparency of the machines energy demand and support an optimised maintenance strategy. The use of a collaborative data pool provides sufficient information to cover a wide range of profitable methods and tools. The software implemented methods and tools enable users with mobile devices opportunities for analysing, controlling and improvement of machine-specific energy demands and maintenance issues on the basis of the same hardware. Further categories of smart environment sensors like rel. humidity, brightness, carbon dioxide, airflow and motion will be introduced.

The distinction between different user roles and their specific information requirements is important to prepare and provide custom-tailored information.

Nevertheless, the introduction of such a system into brownfield requires the consideration of a wide range of subjects. Therefore, it is necessary to achieve acceptance of the employees for the system by supporting their daily work and not to sense that they might get replaced by the machines. Furthermore, topics for IT and security are important, like standardisation of data and interfaces, an extensive authentication management and the encryption and decryption of data. Also, the lack of reliable cost-benefit analysis and difficulties in estimating long-term risks for technical and occupational safety does impede the introduction of the suggested system.

It can be noted that due to a high number of different machine controls and production planning systems, the application of such an approach becomes highly complex.

Next work of the authors will be to extend the methodology to more advanced data mining methods relying on the examination of high-resolution load profiles. Figure 6.2 opens up further possibilities for the analysis of production equipment and the derivation of decision rules. The application of effective feature extraction strategies and the use of proper evaluation models pave the way for reliable prediction on wear out of manufacturing machines and facilities.

The presented approach will be proved within research and industry demonstrators. Hence, a wide range of different manufacturing systems can be considered.

Acknowledgements This paper evolved of the research project ESIMA (improved resource efficiency by power-autonomous sensor systems and customised human–machine interaction) which is funded by the German Ministry of Education and Research (BMBF) within the "Energy self-sufficient mobility—reliable energy self-sufficient systems for the mobile human" research and development programme and managed by the Project Management Agency VDI/VDE IT. Visit http://www.esima-projekt.de/ for more information.

References

Bauernhansl T, Ten Hompel M, Vogel-Heuser B (2014a) Industrie 4.0 in produktion, automatisierung und logistik: anwendung technologien migration. Springer, Berlin, Heidelberg, New York, p 399

Bauernhansl T, Ten Hompel M, Vogel-Heuser B (2014b) Industrie 4.0 in produktion, automatisierung und logistik: anwendung technologien migration. Springer, Berlin, Heidelberg, New York, p 525f

Bogdanski G, Schönemann M, Thiede S, Andrew S, Herrmann C (2013) An extended energy value stream approach applied on the electronics industry. In: Emmanouilidis C, Taisch M, Kiritsis D (eds) Advances in production management systems. Competitive manufacturing for innovative products and services. APMS 2012. IFIP advances in information and communication technology, vol 397. Springer, Berlin, Heidelberg

Bundesministerium für Bildung und Forschung (2018). http://www.bmbf.de/de/9072.php

Chamberlain S, Sharp H, Maiden N (2006) Towards a framework for integrating agile development and user-centred design. In: Marchesi M, Abrahamsson P, Succi G (eds) Extreme programming and agile processes in software engineering. 7th International Conference, XP 2006, Oulu, Finland, 17–22 June 2006. Proceedings 2006. Aufl. Springer, Berlin, Heidelberg, p 143f

DFKI (2014) 6th innovation day of the SmartFactoryKL

DIN EN 13306 (2010–12) Maintenance—Maintenance terminology

DIN EN ISO 9241-210 (2011) p 19

DIN EN ISO 50001 (2011–12) Energy management systems—requirements with guidance for use

Erlach K, Westkämper E (2009) Energiewertstrom – Der Weg zur energieffizienten Fabrik. Fraunhofer Verlag, Stuttgart. ISBN 978-3-8396-0010-8
European Commission (2018). https://ec.europa.eu/clima/policies/strategies_en
Fallenbeck N, Eckert C (2014) IT-Sicherheit und cloud computing; Bauernhansl T, Ten Hompel M, Vogel-Heuser B. Industrie 4.0 in produktion, automatisierung und logistik: anwendung technologien migration. Springer, Berlin, Heidelberg, New York, p 398f
Ganschar O, Gerlach S, Hämmerle M, Krause T, Schlund S (2013) Produktionsarbeit der Zukunft - Industrie 4.0. In: Spath D (ed) IAO, Stuttgart, Fraunhofer, p 56f
Gorecky D, Loskyll M (2014) Mensch-maschine-interaktion im industrie 4.0-zeitalter. In: Bauernhansl T, Ten Hompel M, Vogel-Heuser B (eds) Industrie 4.0 in produktion, automatisierung und logistik: anwendung technologien migration. Springer, Berlin, Heidelberg, New York, p 525; Zamfirescu CB, Pirvu BC, Schlick J, Zühlke D (2013) Preliminary insides for an anthropocentric cyber-physical reference architecture of the smart factory. Stud Inform Control 22(3)
Hermann M, Pentek T, Otto B (2015) Design principles for industrie 4.0 scenarios: a literature review. Dortmund, Technische Universität Dortmund, p 4f
Herrmann Christoph (2010) Ganzheitliches life cycle management, nachhaltigkeit und lebenszyklusorientierung in unternehmen. Springer, Berlin, Heidelberg, New York, p 360f
Kara Sami, Mazhar Muhammad, Kaebernick Hartmut, Ahmed Noor-E-Alam (2005) Determining the reuse potential of components based on life cycle data. CIRP Ann Manuf Technol 54(1):1–4
May G et al (2015) Energy management in production: a novel method to develop key performance indicators for improving energy efficiency. J Appl Energy 149:46–61
Neef B, Schulze C, Herrmann C, Thiede S (2017) Integriertes Energie- und Instandhaltungsmanagement im Kontext Industrie 4.0—Verbesserte Energieeffizienz und Instandhaltung durch Smart Devices und energieautarke kabellose Sensoren, in: Industrie Management 4.0, Ausgabe 1/2017, Energie- und Ressourceneffiziente Produktion
Posselt G et al (2014) Extending energy value stream models by the TBS dimension—applied on a multi product process chain in the railway industry. In: Proceedings of 21st CIRP conference on life cycle engineering 2014, pp 80–85
Posselt G (2015) Towards energy transparent factories. Springer. ISBN 978-3-329-20868-8
Reichel J, Müller G, Mandelartz J (2009) Betriebliche instandhaltung. Springer, Berlin, Heidelberg, New York, p 137f
Seera M, Lim CP, Nahavandi S, Loo CK (2014) Condition monitoring of induction motors: a review and an application of an ensemble of hybrid intelligent models. J Expert Syst Appl 41:4891–4903
Sheikh AK, Al-Sulaiman FA, Baseer MA (2005) Use of electrical power for online monitoring of tool condition. J Mater Process Technol 166:364–371
Thiede S (2018) Environmental sustainability of cyber physical production systems. Procedia CIRP 69:644–649
Tole A (2013) Big data challenges. Database Syst J IV:3:31ff
Umweltbundesamt (2018). https://www.umweltbundesamt.de/daten/energie/energie verbrauch-nach-energietraegern-sektoren
US EPA (2011) Lean, energy & climate toolkit. EPA-100-K-07-003
Ziesemer M (2015) So sieht Industrie 4.0 aus - ZVEI stellt, Referenzarchitektur Industrie 4.0 (RAMI)‘ vor und definiert die, Industrie 4.0 Komponente, in: Technik und Wirtschaft für die deutsche Industrie—Produktion, Nr. 13, p 17

Chapter 7
Two Practical Approaches to Assess the Energy Demand of Production Machines

Eberhard Abele, Philipp Schraml, Martin Beck, Dominik Flum and Christian Eisele

Abstract Energy efficiency becomes an increasingly important quality attribute of modern machine tools. In order to stay competitive and in liability toward our environment, the energy consumption of machine tools must be significantly reduced without deteriorating the productivity and the quality of manufacturing. In line with the rising importance, the number of available energy efficiency solutions is increasing. However, a broad application of these solutions does not exist since the hypothetical saving potentials are hard to evaluate. This chapter presents two self-developed software solutions designed for different application levels. The presented *assessment tool* allows for a quick and easy assessment of the energy demand of a given production machine and is therefore intended for utilization in the procurement process of production machines. In contrast, the *production machine simulation tool* is designed for an assessment of the energy demand in the early product development stages by providing detailed energy and power demand information for all energy-relevant functional modules as well as for the entire production machine. For both approaches, no additional hardware measurements are required as input.

7.1 Introduction

In recent years, the awareness of society, politics, and industry on the subject of energy efficiency has increased. One reason for this is the noticeable environmental impact, which yet again leads to both rising customer awareness and additional legislative regulations. The manufacturing industry, as one of the main consumers of global primary energy and producer of related emissions, represents a great lever to reduce the energy demand worldwide (IEA 2008). Furthermore, energy is an increasingly important cost factor. Within production engineering, the manufacturers

---

E. Abele (✉) · P. Schraml · M. Beck · D. Flum · C. Eisele
Institute of Production Management, Technology and Machine Tools (PTW), TU Darmstadt, Otto-Bernd-Str. 2, Darmstadt, Germany
e-mail: info@ptw.tu-darmstadt.de

S. Thiede and C. Herrmann (eds.), *Eco-Factories of the Future*, Sustainable Production, Life Cycle Engineering and Management,
https://doi.org/10.1007/978-3-319-93730-4_7

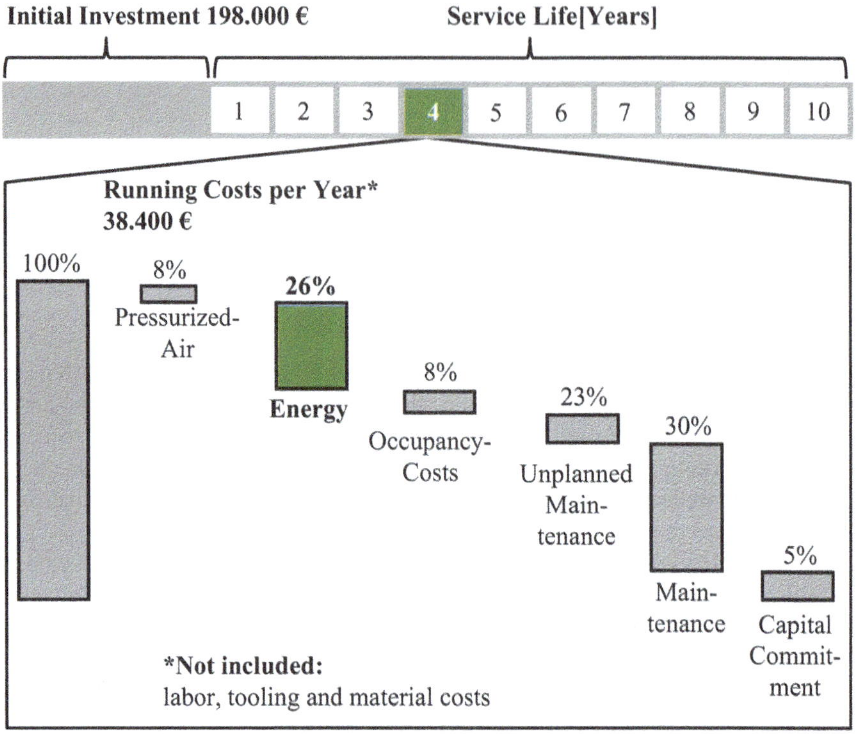

Fig. 7.1 Life cycle costs of an exemplary cutting machine tool (PTW 2008)

of production machines can notice a high customer demand for energy-efficient machines.

Studies have shown that the energy costs for machine tools are already responsible for up to 26% of the running costs—excluding labor, tooling, and material costs (see Fig. 7.1) (PTW 2008). Due to rising energy prices, this share is expected to increase further and consequently the significance of the energy efficiency factor, as one target dimension for production machines besides the classic dimensions precision, power, investment costs, and reliability, will rise (Eisele 2014; BDEW 2013).

7.2 Energy Efficiency of Production Machines

In past research, various approaches have been considered in order to optimize energy efficiency within production engineering. In the context of energy-optimized production machines, the research project "MAXIEM (maximizing the energy efficiency of machine tools)" demonstrates the maximum achievable energy efficiency. Based on the sample four-axis machining center MAG XS 211, various measures for a

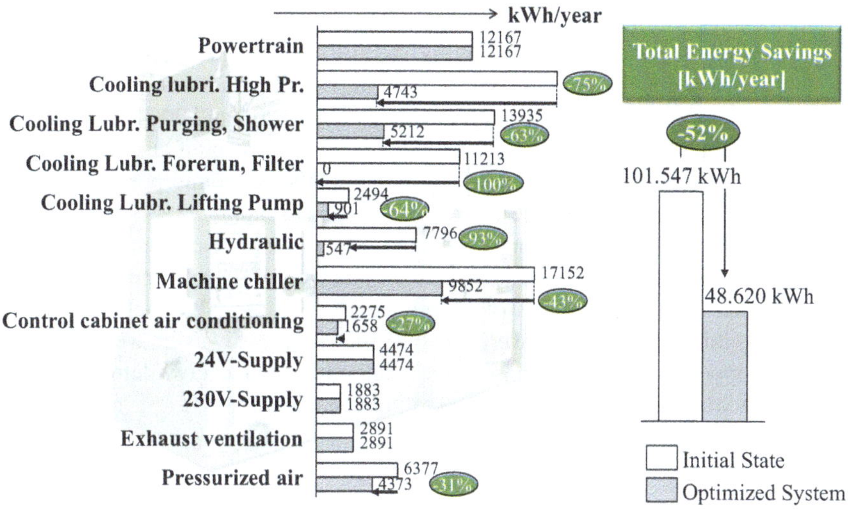

Fig. 7.2 Realized energy-saving measures on a machine tool of the type MAG XS211 (Abele et al. 2012)

component-oriented optimization and evaluation of energy efficiency have been performed. By analyzing the energy consumption of individual components, the actual delivery state of the machining center was determined. The cooling lubricant system, the machine cooling, and the hydraulic system were found to be the most energy-intensive modules. By using an energy-optimized configuration, it could be shown that energy savings of over 50% in relation to the initial state are possible. These results are based on the assumption of a mass production with a three-shift operation and six working days per week (see Fig. 7.2) (Abele et al. 2012).

7.3 Barriers for Implementing Energy Efficiency Measures

As shown in the last chapter, large energy efficiency potentials do exist. However, they are often not seized. This gap between technically and economically feasible measures and the actual implementation in industry is called the "energy efficiency gap" (Jaffe 1994). As possible reasons, *Schmid* identifies three key factors: market failures, organizational weaknesses, and inadequate possibilities for the evaluation of the economic potential (Schmid 2004).

Market failures arise if market participants are not able to include all the necessary information in the decision-making process and so do not use their resources in the most efficient and cost-minimizing way. As an example, a company might not have sufficient information about available energy efficiency measures or could be uncertain about their applicability within the company (Schmid 2004).

Deficiencies within internal company processes and corporate structure can lead to organizational barriers for implementing energy efficiency measures. For example, the energy efficiency factor might not be taken into account in the company's objectives since the focus might be on short payback periods (Schmid 2004). In addition, no tools or methods are currently available to assess the impact of an energy-saving measure. The result is an inadequate evaluation of the economic potential, which prevents investments in energy-efficient technologies (Schmid 2004).

Organizational weaknesses, market failures, and inadequate evaluation possibilities of the economic potential can be ascribed to a lack of transparency concerning energetic and economic impacts of energy-saving measures. Precisely, this transparency would be necessary to overcome the efficiency gap by using well-grounded data. One approach for this purpose is the evaluation of the energy demand of production machines by modelling their operating behavior.

7.4 Modelling the Energy Demand of Production Machines

7.4.1 Theoretical Background

Subsequently, the existing scientific approaches for both the simulation and the approximation of the energy demand of production processes and machines are presented.

7.4.1.1 Process Modelling and Machine Simulation

In the field of cutting machine tools, *Reeber* has performed basic research on energy consumption. He developed an approach which allows for the calculation of the specific cutting energy. He drew on an empirical cutting force model developed by *Victor and Kienzle*, which enabled him to determine the cutting energy depending on various process and material parameters (Reeber 1980; Kienzle and Victor 1957). The calculations do not include the energy consumption of the machine, but only the energy required for the cutting process (Eisele 2014).

Wolfram and *Degner* studied various manufacturing processes in terms of their specific energy demand. They presented an approach that complements the approach of Reeber to the additional energy demand of the machine tool. For this purpose, electrical power measurements were made in the air cut to determine the electrical idle power (all aggregates enabled, no tool contact) (Degner and Herfurth 1983; Degner and Wolfram 1983; Degner et al. 1984; Wolfram 1986, 1990; Eisele 2014).

For the first time, *Schiefer* observed the energy consumption of a machine tool in production-free periods. He assumed a constant power consumption when the machine axes are stationary (Schiefer 2001).

Gutowski developed a similar approach on the basis of electrical power measurements on various machine tools. Additionally, the energy demand of the process-specific portion resulting from the processing load and a base load is calculated. While the base load is assumed to be constant, various milling tests are carried out with varying cutting removal to determine the processing load (Gutowski et al. 2006, 2007).

Draganescu assessed the energy efficiency of machine tools by analyzing the energy effectiveness, which is defined as the ratio of theoretical cutting energy to the total energy demand. By means of statistical methods and empirical data, a mathematical model was developed mapping the relationship of various operating parameters such as spindle speed, torque, and feed rate in order to approximate the specific power consumption of a machine tool (Draganescu et al. 2003; Abele et al. 2015).

7.4.1.2 Machine Modelling and Simulation

Based on graph theory, *Dietmair* used an empirical model to predict the energy demand of cutting machine tools. For every machine component, a certain electric power demand, corresponding to the actual machine mode, was set. By specifying the sequence and the duration of the individual machine modes in the respective observation period, the power consumption and the energy need for the individual components could be concluded. Hence, it was possible to calculate the energy demand of the entire machine (Dietmair et al. 2008; Dietmair et al. 2009; Dietmair and Verl 2009; Dietmair 2010; Eisele 2014).

Bittencourt enlarged the approach of *Dietmair* by assigning a certain time and power demand to the transition between different machine modes. Both, the approaches of *Dietmair* and *Bittencourt,* neglect dynamic effects. Each machine mode is associated with a single power demand (Bittencourt 2013).

On the basis of a dynamic simulation, *Schrems* developed an approach to predict the energy demand of various production machines and processes. The production machines included in a process chain are presented by generic models. By implementing the data sheet disclosures, the energy demand of specific configurations can be assessed. As a result, the energy demand can be taken into account when planning process chains and selecting alternative production machines (Schrems 2014).

7.4.1.3 Energy Demand Approximation of Production Machines

For a prospective assessment of the energy and medium demand, *Kuhrke* developed a methodology that can already be used in the offer phase of machine tools. Therefore, a foundation for machine tool manufacturers as well as for operators for a coherent evaluation of the energy and medium demand is provided. Based on the analysis of a sample machine, he developed calculation rules for each energy-relevant component. In order to achieve this, he used both information from data sheets and data gained by measurements if the required information was not at all or insufficiently available in the data sheets. Finally, the total energy demand of the machine tool can be calculated by aggregating the individual demands (Kuhrke 2011).

In his work, *Rief* presents a predictive model for the energy demand of machine tools. As a modelling base, electric power measurements were made at various modules at different operating conditions of the production machine. For modules with process-dependent power consumption, further series of experiments were carried out at various load conditions. By means of spline curves, a characteristic model was developed which enables the calculation of the energy consumption depending on the production task (Rief 2012).

7.4.1.4 Summary

In contrast to many of the previous works, the approaches presented below do not rely on energy measurements on existing production machines. For one thing, electrical power measurements are often unsuitable in a production environment because they are associated with an increased effort and high costs, since the performance of measurements is time-consuming and the production might need to be stopped. Furthermore, measurements can usually not be used in the planning phase of a product or production due to missing physical components.

The simulation approach illustrated below allows a demand-actuated dimensioning of the modules and the selection of energy-efficient components in the product development phase of production machines. In contrast, the presented assessment tool can especially be used in the procurement process of production machines, providing a quick and easy evaluation of the energy demand of a given production machine.

7.4.2 Quick Scan: Energy Efficiency for Machine Tools

7.4.2.1 Scope of Application

General recommendations for the implementation of specific energy efficiency measures need a dedicated analysis of the use case of a production machine, such as a machine tool. Still, users intend to find appropriate measures enhancing the energy efficiency of their production machines while increasing the profitability simultaneously. For a final user of machine tools, a quick scan is needed to identify key levers. Conventionally, such methods are based on energy measurements of relevant machine tools and their components or on a simulation approach. An easy to applicate approach avoids costly generation and interpretation of energy data, retaining a sufficient accuracy of the obtained recommendations.

Easily accessible data are the basis of a quick scan approach. In existing production environments, the following data sources are of major interest:

- Technical documentation of machine tools,
- Historical production data and future production plans (Thiede 2011), and
- Expert knowledge (Böhner 2013).

In the following, a three-step approach is presented to identify energy efficiency measures and its economic potential:

1. Identification of relevant machines,
2. Definition of energy-relevant components and identification of discretion to act, and
3. Derivation of economic potential of energy efficiency measures.

7.4.2.2 Identification of Relevant Machines

To identify machine tools with a high energy-saving potential, a bundle of information is needed. The operational modes of a producing business small-, medium-, or large-scale production are usually proportional to the load of a production machine and herewith with the load of its components. Information on operating hours of the machine tool is the basis for further investigations. To assess whether saving potentials are rather high or low, the power rating of the machine tool is additional information to be collected.

Operating hours can be derived locally from the production machine by operating hour counters (e.g., spindle hours) or from a central registration of production data, e.g., out of a manufacturing execution system (MES). Information about the power rating can be obtained directly at the machine or is documented in the machine maintenance department.

Beyond this, further data are needed as the energy-saving potential is not necessarily linked to the power rating. These information concerns:

- Main time/secondary time ratio,
- Machine mode outside business hours, and
- Standby management.

To assess these data, a systematic procedure has been developed involving data from technical documentations, historical production data, future production plans, and expert knowledge. As the assessment uses qualitative data, it has turned out to be appropriate to use a relative scale to rate the information gained in the categories described above and thus to obtain priorities for action.

Additionally, the knowledge of the assessing energy expert is indispensably to prioritize, as priorities for action can differ between different production environments. An example of the procedure is shown in Fig. 7.3. The result is a list of the ratings for production machines obtained by the described procedure.

General Machine data				**Rating**
Machine Number		290-06		
Manufacturer?		PTW		
Year of construction		2011		
Residual using time		> 5 years		
Power rating [kVA]		35		7
Is the machine set to a energy efficiency mode (stand-by,swiched off) outside business ours?		no		10
Total operating hours per week		168		10
Shifts (8 h) per day		3 shifts		
working days per week		6 days		
Main time/ secondary time ratio				
Average main time share [%]		45%	35,5 [h]	7
Average secondary time share [%]		55%	43,4 [h]	
	Number of work pieces produced on the machine per week	**machining time [min]**		
1st work piece	500	2	16,7 [h]	
2nd work piece	280	3,5	16,3 [h]	
3rd work piece	230	6	23,0 [h]	
4th work piece	170	5	14,2 [h]	
5th work piece	150	3,5	8,8 [h]	
		Total	78,9 [h]	
Non working time within operation hours per week [h]		53%	89,1 [h]	7
Total rating				8,7

Fig. 7.3 Identification of energy efficiency-relevant machine tools

7.4.2.3 Definition of Energy-Relevant Components and Identification of the Potential for Action

The starting point in order to identify relevant machine components is the determination of a component condition matrix. Predefined energy-relevant components of the machine tool have to be analyzed regarding their state of operation in different machine modes. To define energy-relevant components, information about the machine tool structure and expert knowledge is essential. An established classification (VDMA 2014) distinguishes the following machine modes:

- Working
- Operational
- Powering up
- Standby
- Off.

It has turned out to be appropriate to differentiate the mode operational further in setup/tool change and in waiting. The mode powering up normally can be aggregated usually to the working mode.

Machine mode \ Component	Main Drives	HMI	Cutting lubricant pump (high pressure)	Cutting lubricant pump (low pressure)	Hydraulics	(…)
Working	On	On	On	On	On	On
Operational	On	On	On	On	On	On
Waiting	On	On	On	On	On	On
Stand-by	Off	On	Off	Off	Off	Off
Powering up	On	On	On	On	On	On

Fig. 7.4 Identification of relevant components based on the component condition matrix

Based on the derived timeshares, the state of operation of the machine tool component has to be determined. Herewith, a matrix for each energy-relevant component can be created (according to Fig. 7.4). Of prior interest are components which are not operating need-based. These are components which are running even if the machine is not set to the working mode, while there is no need for specific useful energy. This may be the exhaust system, the cooling fan of a decentral cooling device of a machine tool, pumps in the cooling lubricant system, or the chip conveyor.

In the following, the overall using time of each component can be calculated and the energy demand of each energy mode can be assessed. Therefore, the working mode (e.g., pressure and volume flow) as well as information about the efficiency level in the nominal operating point of the component is necessary. Additionally, appropriate assumptions have to be made which (part load) operating modes the component can be set.

The same procedure can be conducted with the technical alternatives. The result is a comparison of the prevailing component setup with an energy-efficient setup and the assigned overall energy demand of the setup alternatives derived from the methodology described and shown in Fig. 7.5.

To derive the economic potential of different energy-efficient component configurations, the average energy price has to be set as well as the target amortization time. Based on a static investment calculation, the maximal investment amount can be derived. This amount can be the basis for negotiations with component suppliers.

Cutting lubricant pump (high pressure)					
Operational	Working	4106	[h/yr.]	switched on	
	Main time	1848	[h/yr.]	switched on	
	Secondary time	2258	[h/yr.]	switched on	
	Set-up / tool change	1198	[h/yr.]	switched on	
	Waiting	2172	[h/yr.]	switched on	
	Stand-by	0,00	[h/yr.]	switched off	
Time share of cl usage during chip removal		65%			
Share small tools		60%			
Share medium tools		25%			
Share big tools		15%			
Pressure regulation concept		pressure control valve			
Maximum pressure		50,00	[bar]		
Volume flow		38,00	[L/min]		
Overall efficiency of motor-pump-system		75%			
Hydraulic capacity		3,17	[kW]		
Electrical power demand		4,22	[kW]		
		Usage during chip removal	Needed hydraulic capacity	Electric power demand: Current	Electric power demand: Optimized
Time shares	small tools	9,6%	60%	4,2 kW	2,53 kW
	medium tools	4,0%	80%	4,2 kW	3,38 kW
	big tools	2,4%	90%	4,2 kW	3,80 kW
	pressure control valve without chipping	83,9%	100%	4,2 kW	0,00 kW
Energy demand standard system		48.899	[kWh /yr.]		
Energy demand optimized system		5.459	[kWh /yr.]		
Engergy costs		0,15	[€/ kWh]		
Target amortization time		2	[yr.]		
Energy saving potential		43.439	[kWh /yr.]		
Max. recommended investment		13.032	[€]		

Fig. 7.5 Derivation of the economic potential of energy efficiency measures

7.4.3 Energetic Simulation of Production Machines

A popular approach in order to predict the behavior of production machines is the use of simulation models. According to VDI3633, simulation can generally be applied for technical systems within every life cycle phase: During product development, the predicted system behavior can be verified; in the use phase, potential changes to the utilization profile or retrofit measures can be evaluated in advance (VDI 2013; Abele et al. 2015).

In this context, the aim of the presented work is the development of a simulation-based assessment for the energy demand of production machines. Through the use of the developed simulation approach, production machine manufacturers as well as operators should be enabled to assess the energy demand of a production machine in an easy and standardized way. With minimal input data and most importantly no additional hardware power measurements, maximal prediction accuracy is to be achieved. Through modelling all components, relevant to the energy demand, and the energetic interconnectivity, all functional modules of a production machine are set up. Besides electrical energy, all other energy types are taken into account influencing the electric behavior (e.g., hydraulic and mechanical energy). The simulation model is implemented within the software environment MATLAB/Simscape.

The presented simulation approach covers the relevant mechanisms for calculating the energy demand of production machines. In general, the simulation structure distinguishes between three model layers (compare to Fig. 7.6) (Eisele 2014):

- **Process layer (NC code interpreter)**: Within the process layer, the interaction between the workpiece and tool is mapped. Cutting force calculations, as well as tool engagement estimations, are performed in this part of the simulation model. By transforming the calculated cutting forces into torque on the main spindle as well as forces on the feed drives, the load on the drive system of the production machine can be predicted.
- **Machine layer (machine simulation)**: Every functional module relevant to the energy demand of a production machine is mapped in the machine layer. Depending on the available data sheet information of the component manufacturer, physical/mathematical interrelationships and characteristic curves/maps are used for setting up the simulation models. Besides fixed component parameters, the model behavior depends on dynamic interactions on functional module level, process-dependent loads from the process layer, and control commands generated by the control layer.
- **Control layer (model control)**: The control layer represents the physical machine control of a production machine. Using input data from the process layer set, speeds for feed and spindle drives, as well as switching information of peripheral systems, are provided to the machine layer.

7.4.3.1 NC Code Interpreter and Model Control

The basic functionality a production machine has to offer is the desired value-adding process. All peripheral systems of a production machine support and depend on this aim. Thus, a detailed production machine simulation has to include a machining model since it builds the basis for the behavior of almost all other functional module simulation models. The developed NC code interpreter module consists of two parts (compare to Fig. 7.7). Within the *interpreter part* DIN66025 conform parts programs can be evaluated regarding

- **Switching commands**: Time or event-triggered control commands leading to binary on/off operations of functional modules or components of functional modules. An example of a time trigger command is the activation/deactivation of the

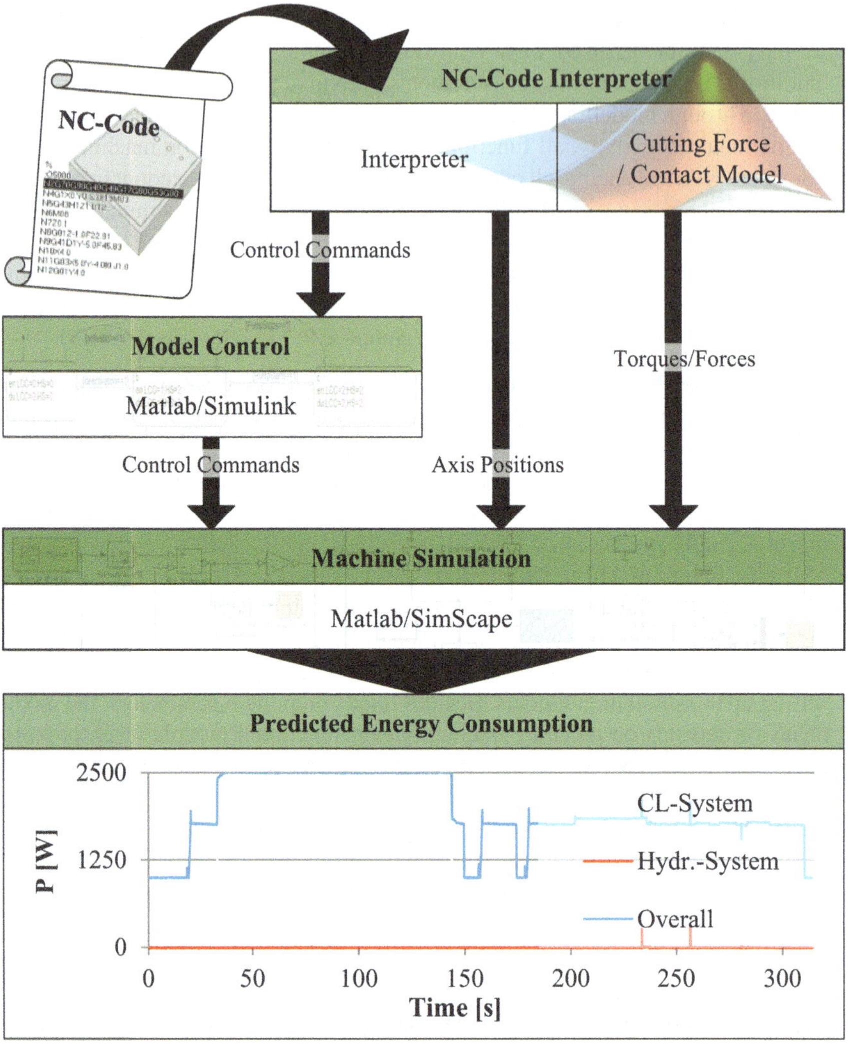

Fig. 7.6 General simulation structure

high-pressure coolant lubricant system. In contrast, a programmed tool change can lead to several event-triggered events.

- **Technology data**: Machining relevant information such as spindle speed, feed rate, and tool number.
- **Movement data**: Position data of tool and workpiece.

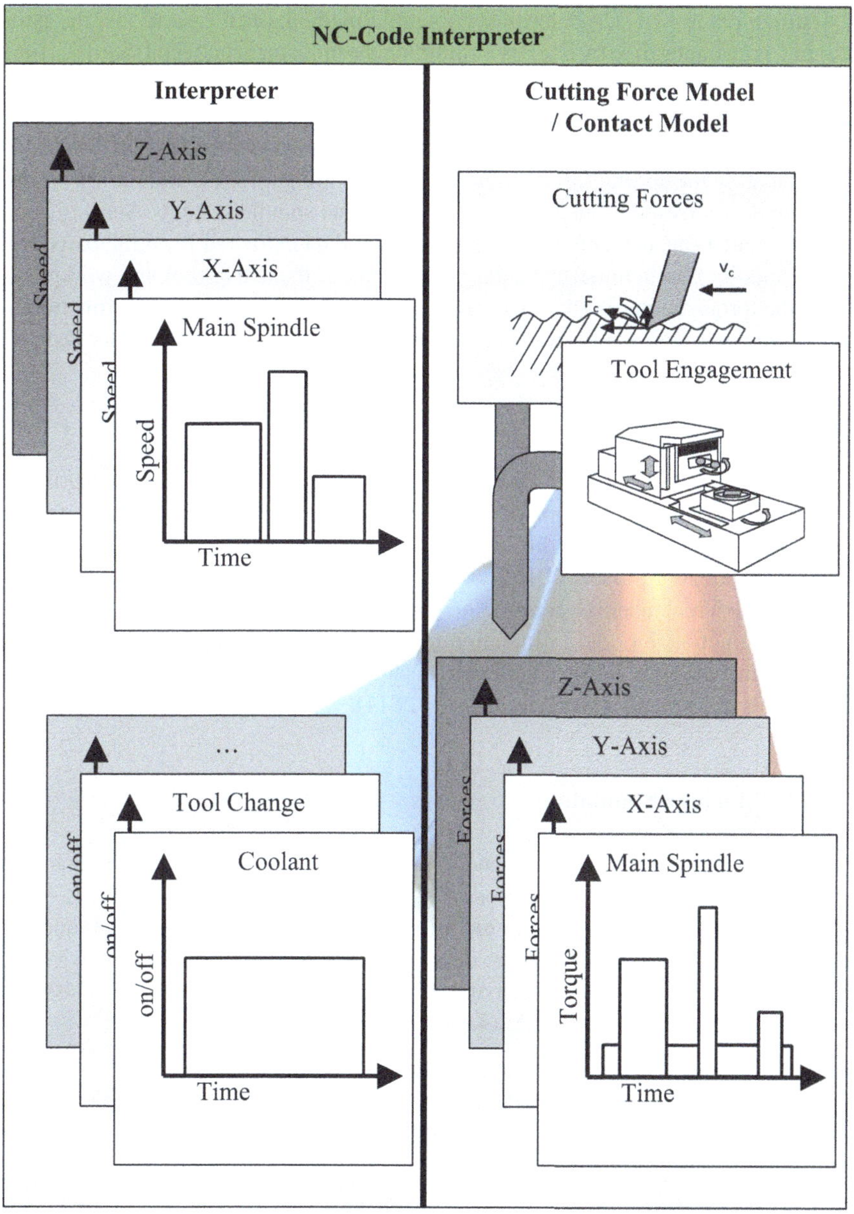

Fig. 7.7 Developed NC code interpreter

Using a plane MATLAB script with simple mathematical operations, the parts program is evaluated line by line. By combining the movement and feed rate data, time and event (axis position, switching commands, etc.) vectors are generated through interpolation.

The *cutting force/contact model* of the NC code interpreter module is responsible for calculating the cutting forces related to the machining process and for transforming them into corresponding loads on feed drives and spindle systems. Starting point for the calculations is technology and movement data extracted from the parts program in addition to geometrical and material data of the active tool and workpiece. Thus, the torque acting on the main spindle M_c can be determined as a function of the specific cutting force F_c, number of teeth in mesh z_e, and tool diameter D_{tl}:

$$M_c = F_c * z_e * \frac{D_{tl}}{2} \tag{7.1}$$

The specific cutting force F_c is obtained using the empirical model of *Victor and Kienzle* and the model extension of *Pauksch* including correcting values, e.g., for the use of coolant lubricant and different cutting materials (Kienzle and Victor 1957; Paucksch et al. 2008; Eisele 2014).

Similar to a real machine control, the model control is responsible for controlling the single functional module simulation models. Taking the extracted NC code data as input, the model control enables or disables modules, and sets limits to axis movements and spindle speed, etc. (Eisele 2014).

7.4.3.2 Machine Simulation

The main driver for the energy demand of production machines is auxiliary functions supporting the actual machining process (Abele et al. 2012). Thus, one focus of the presented simulation approach lies on the exact modelling of all functional modules relevant to the overall electric energy demand. In general, the operating behavior and thus the corresponding energy demand of all functional modules can be classified into the following three categories (Kuhrke 2011):

- Constant demand—e.g. 24-V supply and machine control,
- Cyclic demand—e.g. chiller systems and lift pumps, and
- Variable demand—e.g. main drives.

Corresponding to the operating behavior, the single functional module simulation models are set up. The input variables for the simulation model are the extracted switching commands, technology and movement data, as well as physical parameters taken from the data sheets provided by the component manufacturers (e.g. motor inductivities, characteristic curves of fluid pumps, etc.). Depending upon the available information from the modelled component data sheet, the simulation models vary.

If detailed characteristic curves/maps, representing the components behavior, are available, they are transformed into MATLAB readable lookup tables and directly

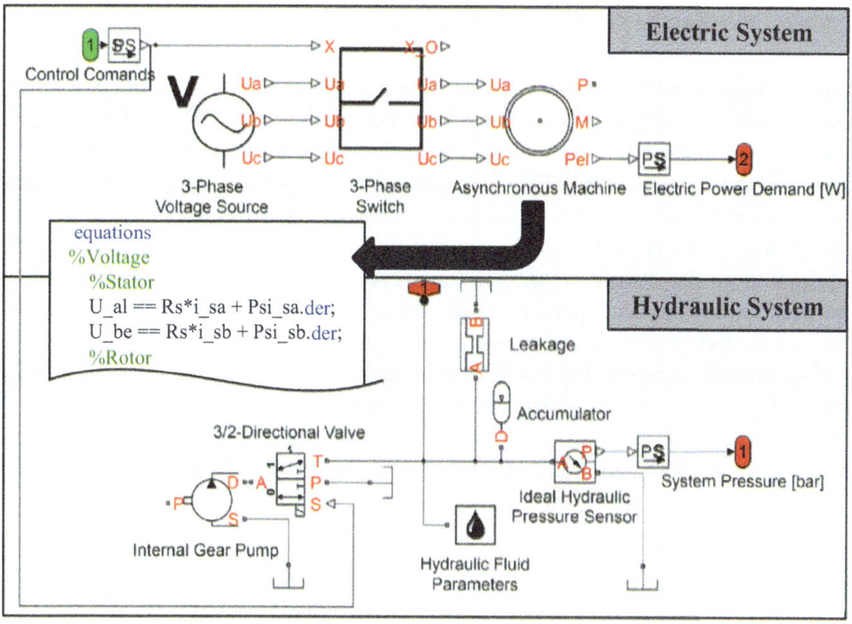

Fig. 7.8 MATLAB/Simscape model of a simple hydraulic system

used for a simulation. A common example is fixed displacement fluid pumps where often characteristic curves illustrating the relationship between acting pressure and required mechanical energy exist.

For other functional module components, such information is not available. In these cases, the component behavior is modelled using their representing physical/mathematical model. This is the case, e.g., for all electric drives.

The functional module models are implemented within the simulation environment MATLAB/Simscape. Simscape, as an extension of MATLAB/Simulink, is software for dynamic simulation of technical systems. In contrast to plain Simulink, Simscape allows modelling of physical components using an independent programming language. With its object-oriented approach, it is easy to build simulation model libraries. Simscape offers the possibility to integrate commercial MathWorks libraries for all kinds of technical systems (mechanics, hydraulics, power systems, etc.). Since those libraries are connected to additional software license costs and do not offer the needed flexibility for adaption, all simulation models on component level are developed independently. Figure 7.8 exemplarily displays a Simscape simulation model of a simple hydraulic system.

By merging and interlinking the developed component simulation models to functional modules, production machines can be mapped successively (Eisele 2014).

7.4.3.3 Exemplary Simulation Application

During the project EMC², the developed simulation approach was applied to a five-axis machining center which was used for titanium alloy machining in the aerospace industry. The machined part was a structural element of an aircraft belly with dimensions in length exceeding width and height by far. The machining process itself involved a multitude of different machining operations with a total duration of about 2 h. Within a first step, the main electric energy-consuming functional modules of the production machine were mapped in the previously presented simulation environment. Based on the actual parts program, the power/energy demands for all relevant functional modules, as well as the entire machine tool, were predicted.

The simulation results for a dedicated finishing operation with a face milling tool are displayed in Fig. 7.9. While the graphical results of the contact model are shown in *a*, simulated as well as measured power demand of the total machine tool is presented in *b*. During the first seconds, the drive system as well as the coolant lubricant system is switched on. The tool and workpiece engagement occur at about 22 s (simulation 8 s). The load jump between 190 s and 470 s (simulation 195 s and 450 s) is due to the chiller system of the machine tool. At 505 s (simulation 510 s), the tool retracts from the workpiece. The relative deviation between measured and simulated energy demand is less than 5% and is mainly due to differences between real occurring cutting forces and the simulated cutting forces based on an empirical model. Switching to a more advanced model could improve the accuracy. The other obvious deviation is the difference in length of the load jump induced by the chiller system. Since the chiller model is based on information publicly available from the manufacturer's data sheet, such variances can occur. By adapting the chiller parameters (e.g., real cooling capacity), the simulation results can be further enhanced.

7.5 Summary and Outlook

Due to the rising share of energy efficiency solutions available on the market, it becomes even harder to evaluate the hypothetical saving potential for a given application. Thus, there is a lack of information preventing a broad application of energy efficiency solutions. Within this chapter, several approaches to reduce this information asymmetry for production machines are presented. The chapter especially focuses on two self-developed software solutions designed for different application levels. While the presented assessment tool is designed for an application within the procurement process of production machines, allowing a quick and easy assessment of the energy demand of a given production machine, the production machine simulation tool can give detailed energy and power demand information for all relevant functional modules as well as the entire production machine. Thus, the application of the production machine simulation tool is especially intended for the machine design process. Both approaches have in common that no additional hardware mea-

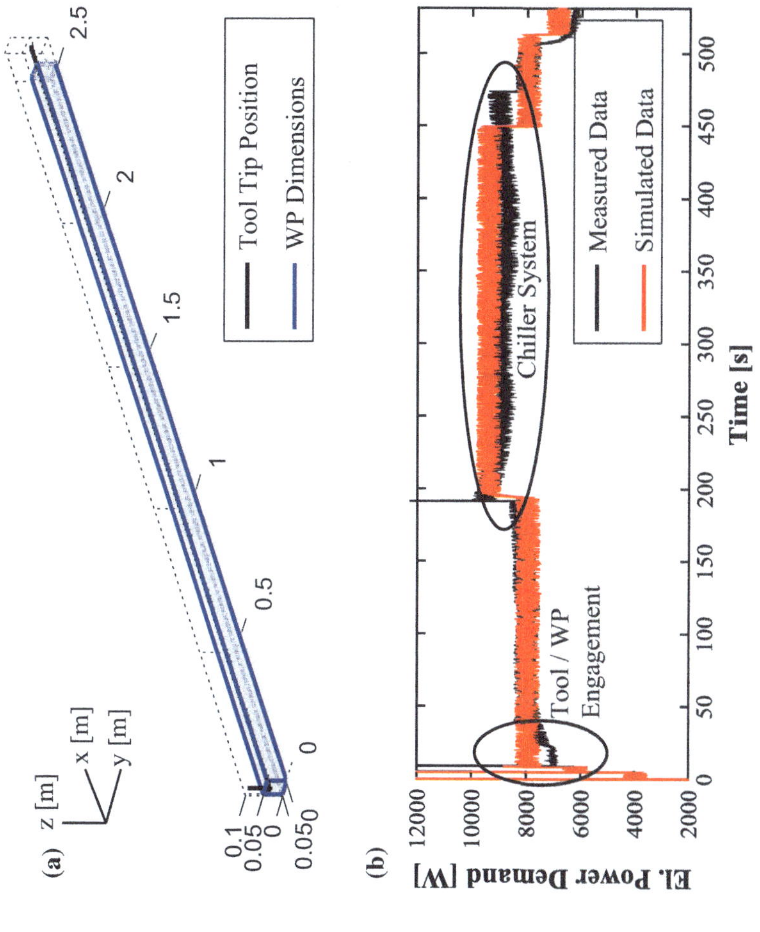

Fig. 7.9 **a** Graphical results contact model and **b** comparison of simulated versus measured power demand

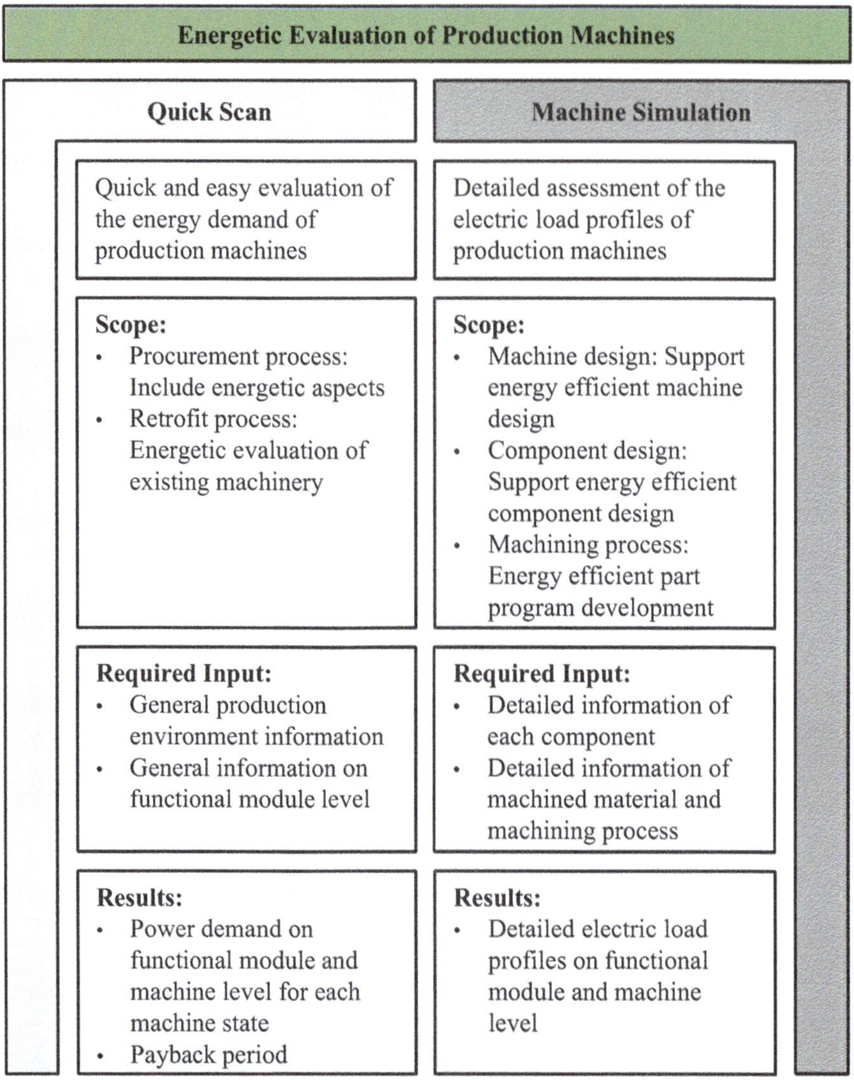

Fig. 7.10 Comparison of the two presented production machine evaluation tools

surements are required as input information. A general overview of the two presented approaches is displayed in Fig. 7.10.

While both presented approaches focus on machine and submachine level, a future application could be the simulation of an entire factory building, not only to evaluate single energy efficiency measures but to have a tool for supporting the design process of energy networks on process chain as well as factory level. For this purpose, the presented production machine simulation models need to be extended with adequate model interfaces to be connected to technical building services and factory building simulation models. By doing so, all relevant electrical and thermal energy flows within a factory can be mapped and factory system designs are tested.

Acknowledgements The research leading to these results has received funding from the European Community's Seventh Framework Program (FP7/2007-2013) under grant agreement n° 285363. The funded project's title is "Eco Manufactured Transportation Means from Clean and Competitive Factory."

References

Abele E, Sielaff T, Beck M (2012) Konfiguration energieeffizienter Werkzeugmaschinen. Werkstattstechnik online: wt 102:292–298

Abele E, Braun S, Schraml P (2015) Holistic simulation environment for energy consumption prediction of machine tools. In: 22nd CIRP conference on life cycle engineering in Sydney, Australia, 7–9 Apr 2015. Paper Accepted

BDEW Bundesverband der Energie- und Wasserwirtschaft (2013) Energiewirtschaftliche Entwicklung in Deutschland. 1. Quartal 2013, Berlin

Bittencourt JL (2013) Selbstoptimierende und bedarfsgerechte Steuerungsstrategien für Werkzeugmaschinen zur Steigerung der Energieeffizienz. Apprimus, Aachen

Böhner J (2013) Ein Beitrag zur Energieeffizienzsteigerung in der Stückgutproduktion. Shaker Verlag, Aachen

Degner W, Herfurth K (1983) Energieaufwand bei spanender Teilefertigung. Fertigungstechnik und Betrieb 33:684–687

Degner W, Wolfram F (1983) Richtwerte und Regeln für den Energieaufwand bei spanender Fertigung. Fertigungstechnik und Betrieb 33:739–742

Degner W, Resch R, Wolfram F (1984) Energieaufwand spanender Fertigungsverfahren und das Prinzip der vergegenständlichten Energie in der Teilefertigung. Metallverarbeitung 83:132–134

Dietmair A, Verl A, Wosnik M (2008) Zustandsbasierte Energieverbrauchsprofile – Eine Methode zur effizienten Erfassung des Energieverbrauchs von Produktionsmaschinen. wt Werkstattstechnik online 98:640–645

Dietmair A, Verl A (2009) Energy consumption forecasting and optimisation for tool machines. Mod Mach (MM) Sci J, Paper 4

Dietmair A, Verl A, Huf A (2009) Automatisierung spart Energie – Direkte und indirekte Maßnahmen in der Gerätetechnik. In: Energy 2.0, p 26

Dietmair A (2010) Energy consumption assessment and optimisation in the design and use phase of machine tools. In The International Academy for Production Engineering (CIRP) (ed): Proceedings of the 17th CIRP international conference on life cycle engineering

Draganescu F, Gheorghe M, Doicin C (2003) Models of machine tool efficiency and specific consumed energy. J Mater Process Technol 141:9–15

Eisele C (2014) Simulationsgestützte Optimierung des elektrischen Energiebedarfs spanender Werkzeugmaschinen. Shaker Verlag, Aachen

Gutowski T, Dahmus J, Thiriez A (2006) Electrical energy requirements for manufacturing processes. In The International Academy for Production Engineering (CIRP) (ed): Proceedings of the 13th CIRP international conference on life cycle engineering, pp 623–627

Gutowski T et al (2007) A thermodynamic characterization of manufacturing processes. In International Association of electronics recyclers (ed): Proceedings of the 2007 IEEE international symposium on electronics & environment, pp 137–142

IEA International Energy Agency (2008) Worldwide trends in energy use and efficiency

Jaffe AB (1994) The energy-efficiency gap. What does it mean? Energy Policy 22:804–810 (Elsevier, Amsterdam)
Kienzle O, Victor H (1957) Spezifische Schnittkräfte bei der Metallbearbeitung. Werkstofftechnik und Maschinenbau 47:224–225
Kuhrke B (2011) Methode zur Energie- und Medienbedarfsbewertung spanender Werkzeugmaschinen. epubli, Berlin
Paucksch E et al (2008) Zerspantechnik: Prozesse, Werkzeuge. Technologien. Vieweg+Teubner, Wiesbaden
PTW TU Darmstadt (2008) CO$TRA—Life Cycle Costs Transparent, Abschlussbericht – Kurzfassung
Reeber R (1980) Der Energiebedarf bei trennenden Fertigungsverfahren. Werkstatt und Betrieb 113:109–113
Rief M (2012) Vorhersagemodell für den Energiebedarf bei der spanenden Bearbeitung für eine energieeffiziente Prozessgestaltung. Shaker, Aachen
Schiefer E (2001) Ökologische Bilanzierung von Bauteilen für die Entwicklung umweltgerechter Produkte am Beispiel spanender Fertigungsverfahren. Darmstädter Forschungsberichte für Konstruktion und Fertigung
Schmid C (2004) Energieeffizienz in Unternehmen: eine wissensbasierte Analyse von Einflussfaktoren und Instrumenten. vdf – Hochschulverlag AG, Zürich
Schrems S (2014) Methode zur modellbasierten Integration des maschinenbezogenen Energiebedarfs in die Produktionsplanung. Shaker Verlag, Aachen
Thiede S (2011) Energy efficiency in manufacturing systems. Springer, Berlin
VDI Verein Deutscher Ingenieure (2013) Simulation von Logistik-, Materialfluss- und Produktionssystemen (3633). Beuth Verlag GmbH, Berlin
VDMA Verband Deutscher Maschinen- und Anlagenbauer (2014) Messvorschrift zur Bestimmung des Energie- und Medienbedarfs von Werkzeugmaschinen in der Serienfertigung, Entwurf (34197). Beuth Verlag GmbH, Berlin
Wolfram F (1986) Aspekte der energetischen Bewertung von Produkten und Prozessen der Abtrenntechnik nach dem Prinzip der vergegenständlichen Energie. Dissertation, Karl-Marx-Stadt
Wolfram F (1990) Energetische produktbezogene Bewertung von Fertigungsprozessen. Dissertation, Chemnitz

Chapter 8
Analysis of Production Lines with Switch-Off/On Controlled Machines

Nicla Frigerio and Andrea Matta

Abstract The implementation of control strategies that reduce energy demand during the machine idle periods is becoming a challenging goal to achieve energy efficiency in production systems. A general policy for switching the machine *off/on* has been recently proposed in the literature for single machines when a transitory is needed to resume the service. This work analyzes the performance of a production line when a control policy is applied at machine level. In more details, the control of one machine at the time and the control of all machines simultaneously are compared. The considered performance measures are the total energy demanded per part and the system throughput. Numerical results are based on discrete event simulation, and a comparison with the most common practices in manufacturing is also reported.

8.1 Introduction

In the last years, energy efficiency in manufacturing is becoming a challenging goal due to the demand of this sector in the worldwide scenario. Indeed, energy saving in production plants is becoming more and more important to contain the environmental impact of manufacturing, and, nevertheless, to reduce costs. One of the most discussed technical solutions to reduce energy waste in production systems is the implementation of energy control strategies during the machine idle periods. This work studies the performance of a production line when a state control strategy is applied at machine level toward the improvement of energy efficiency.

N. Frigerio (✉) · A. Matta
Department of Mechanical Engineering, Politecnico di Milano,Milan, Italy
e-mail: nicla.frigerio@polimi.it

A. Matta ·
e-mail: andrea.matta@polimi.it

S. Thiede and C. Herrmann (eds.), *Eco-Factories of the Future*, Sustainable Production, Life Cycle Engineering and Management,
https://doi.org/10.1007/978-3-319-93730-4_8

8.1.1 Literature Review

The power requirement of a machine tool can be divided into two main components. A *Fixed Power*, demanded for the operational readiness of the machine and independent from the process, and a *Load Dependent Power*, demanded to distinctively operate components enabling and executing the main process (Dahmus and Gutowski 2010; Gutowski et al. 2009). Indeed, the machine auxiliary equipment keeps consuming energy during machine nonproductive periods. This generates a supply excess that could be reduced by controlling the machine state. As a consequence, the energy required by a machine tool can be significantly reduced applying a control at machine level that aims at reducing the machine fixed power demand, which is required even if the production is not requested. Recently, the potential of machine state control during nonproductive periods has been highlighted by several researchers. To give some examples, Ghadimi et al. estimated up to 26% of energy consumption savings in the analyzed scalping process line by applying a switch-off policy (Ghadimi et al. 2014). Weinert and Mose simulated different production scenarios where advanced standby strategies are implemented on production equipment to save energy during short interruptions (Weinert and Mose 2014). The potential saving is shown to be up to 53% of the energy consumed in nonproductive periods. Langer et al. showed the effectiveness of their energy sensitive manufacturing execution system (MES) using a simulation model (Langer et al. 2014). The energy control strategies applied reached significant savings (16–24%) during the nonproductive periods. The use of simulation which allows an energy simulation of machines under several production scenarios is a powerful method to analyze complex systems. Among the others, Abele et al. proposed a methodology to parametrize and simulate the energy behavior of a machine tool according to its state (Abele et al. 2012). Heinemann et al. proposed a hierarchical evaluation tool to analyze the impact at different level of several energy measures, e.g., changing lot size, equipment, process parameters (Heinemann et al. 2014).

Currently, most of the machine tools do not have eco-green functionalities, and in the industrial market there are only a few energy-saving control systems available. Most of them have been developed by machine tool builders in order to support the final users. As examples among the existing options in the industrial market, some companies provide devices for shutdown the machine tool, or some functional modules, once the machine idle period exceeds a user-defined limit (DMG Energysave 2018; Heidenhain 2011a, b). However, the selection of the policy parameters is not supported by any tool or method and is often experience based.

In the last years, several research efforts focused on controlling production systems by switching *off* and *on* the machines to minimize total energy demand when a certain transitory is needed to resume the service. Mouzon et al. presented several switch-off dispatching rules for a non-bottleneck machine in a job shop (Mouzon et al. 2007). Chen et al. formulated a constrained optimization problem for scheduling machines into *on* and *off* modes in a production line (Chen et al. 2011, 2013). Sun and Li proposed an algorithm to estimate opportunity windows for real-time energy control in a machining line (Sun and Li 2013). Chang et al. analyzed several real-time machine switching strategies using energy-saving opportunity windows in a machining line under random failures, but no start-up time is considered (Chang et al. 2013). The same approach has been applied to an integrated serial production line with HVAC (Heating, Ventilating, and Air Conditioning) system (Brundage et al. 2014). Mashahei and Lennartson proposed a control policy to switch-off machine tools in a pallet constrained flow shop (Mashaei and Lennartson 2013). This policy aims at minimizing the energy demand under machine random failures where the processing time and the start-up time are deterministic and constant. Frigerio and Matta studied analytically the switching policy for single machines under the assumption of stochastic arrivals, constant start-up, and no information from the buffer in front of the machine (Frigerio and Matta 2015). Under quite general assumptions, they showed that the switch-off and switch-on problems are independent. Therefore, it is possible to independently find the optimal control for switch-off and the optimal control for switch-on. An efficient algorithm has been proposed to find the optimal control—i.e., when to switch-off and -on—considering also two constraints related to production performance. This algorithm also requires less computational effort yielding to a more efficient resolution. Moreover, the authors analytically demonstrated that the optimal policy degenerates into simpler policies for increasing and decreasing hazard rate distributions. In a different work, they modeled explicitly the start-up time as dependent on the time period the machine stays in a low power demand state (Frigerio and Matta 2014b).

Another stream of research related to the machine state control during nonproductive periods is the queueing theory where machines may become temporarily unavailable. The machine vacation usually starts at the completion of a busy period—i.e., exhaustive vacation—and the service is resumed according to some policies. The simplest resumption policies are the *N-policy* and the *T-policy*. The *N-policy* resumes the service when the queue length accumulates N parts, whereas the *T-policy* resumes the service according to a timer that counts down T time units after a certain event. The combination of the two policies leads to the hybrid *NT-policy*. Detailed surveys have been reported by Tian and Zhang (2006), Ke et al. (2010) and others. Of particular interest is the work of Yadin and Naor 1963, who proposed firstly the *N-policy* in a queueing system where a random transitory is needed to resume the service.

In computer science, this theory has been applied in order to study the trade-off between energy efficiency and mean delay in cellular network (e.g., Maccio and Down 2013) and in server farms (e.g., Guo et al. 2013), when the buffer capacity is assumed to be infinite.

In the manufacturing field, Frigerio and Matta (Frigerio and Matta 2014a) proposed and analyzed an hybrid control policy—i.e., the *TNT-policy*—for machine tools using buffer information. The closedown time is a control parameter (τ_{off}), as well as a threshold on buffer level (N) and the vacation duration ($\tau_{\text{on}} - \tau_{\text{off}}$). The control parameters that minimize the machine expected energy demand are provided numerically for an M/M/1/K system with exponential start-up duration.

8.1.2 Objectives and Contributions

From the literature analysis, it emerges that the energy reduction problem has been studied at local level when a single machine can be controlled using local information, such as arrivals and process time information. However, as far as authors knowledge, the impact of controlling one machine on production system performance has never been discussed yet. Moreover, at system level, some switching control strategies have been proposed in the literature, but the structural properties of the optimal policy have never been investigated. For instance, it is not known which machines, in a manufacturing system, have to be controlled and which not, and how a simultaneous control of the machines may affect system performance. Indeed, nowadays in most of the manufacturing systems there is a lack of data in terms of energy demand and about the savings achievable by properly controlling the production equipment. Moreover, blocking and starvation can be caused by failures, long processing times and process variability and they are known as symmetrical phenomena. Indeed, blocking propagates upstream because of finite buffer capacity, whereas starvation propagates downstream. However, the machines downstream the failure that are in starvation can be switched off, while machines upstream may have to be kept on while blocked because the part is loaded on the machine (Fig. 8.1). Therefore, a control may create asymmetry.

This work studies the optimal control of a single server queueing system with general arrival distribution. Particularly, the performance of a finite buffer production line is analyzed when a general control policy is applied at machine level. Discrete event simulation is used for the performance evaluation, with an ad-hoc template built in ARENA© software environment for modeling a general machine controlled with an energy state control policy. Several simulation experiments are presented in this work and a comparison with the common practice in manufacturing is also reported. Moreover, in the study, a real CNC machining center is considered, experimentally characterized to estimate the power demand in its different states.

Production lines are chosen as the subject of the analysis because switching policies may have a high impact in terms of machine nonproductive times.

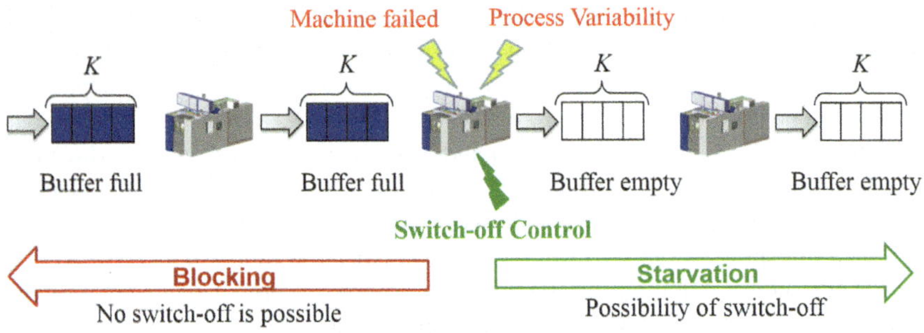

Fig. 8.1 Blocking and starvation phenomena in controlled production systems

Briefly, it will be showed that:

- A general switching off/on policy allows to reach high energy saving without largely compromising the production throughput;
- Locating controlling machines at the beginning or at the end of a balanced production line makes a difference;
- In balanced production lines where each machine is controlled, the optimal parameter values of the policy have a special, and nice, structure that can be profitably used to identify the best system control strategy.

Since these results were obtained with simulation, there is no claim that they are general and valid for any production line. However, this is the first study in which a general policy is analyzed on a production line, and the considerations extracted from this work will be helpful for many researchers and practitioners.

8.2 Assumption

A serial production line with m workstations, each composed by a finite input buffer and a single machine working a single part type, is considered as the system to be controlled (Fig. 8.2). This assumption is valid for lines specialized for one single part type or for a family of similar items, and for lines working large batches while considering a single batch. However, since the production system is assumed to have one single part type, the arrival process is not renewable in general (Wu and McGinnis 2013).

The notation used in this work is as in Table 8.1.

At any station, a part is sent to the machine only during its idle periods, otherwise the part has to wait in the queue. After the completion of the process, the part leaves the station and enters in the downstream buffer. If there is no capacity to hold the part in the buffer downstream, the part is stacked in the machine and the machine itself is blocked—i.e., blocking after service. We assume the machine cannot be switched off when is blocked. The control aims at selecting the optimal system configuration so that the energy demand is minimal.

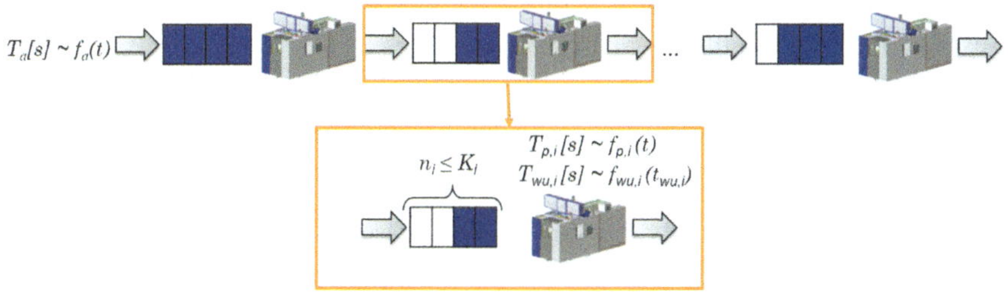

Fig. 8.2 Production line

Table 8.1 Notation

n_i	Number of parts in station i	[parts]
K_i	Capacity of buffer i	[parts]
$\mathscr{S}_i$	Station i state set	–
$\mathscr{S}_X$	Machine state set	–
$T_a \sim f_a(t)$	Arrival time and PDF	[s]
$\bar{t}_a$	Mean interarrival time	[s]
$t_{d,i}$	Time instant of last departure at machine i	[s]
$T_{p,i} \sim f_{p,i}(t)$	Processing time of machine i and PDF	[s]
$\bar{t}_{p,i}$	Mean processing time of machine i	[s]
$T_{\text{su},i} \sim f_{\text{su},i}(t)$	Start-up duration of machine i and PDF	[s]
$\bar{t}_{\text{su},i}$	Mean start-up duration of machine i	[s]
x_i	Energy state of machine i	–
$w_{x,i}$	Power demand of machine i in state x_i	[kW]
$w_{q,i}$	Holding power demand per part of buffer i	[kW]
y_i	State of station i (i.e., machine + buffer)	–
α_{TNT}	Control policy	–
$\mathbf{c}_i$	Control parameters for machine i	–
N_i	Buffer level to switch-on the machine i	[parts]
$\tau_{\text{off},i}$	Time interval to switch-off after the last departure from machine i	[s]
$\tau_{\text{on},i}$	Time interval to switch-on after the last departure from machine i	[s]
$\mathbb{E}[P_i]$	Expected power demand of station i	[kW]
$\mathbb{E}[Q_i]$	Expected energy demand of station i	[kJ/s]
$\mathbb{E}[Q_{\text{tot}}]$	Total expected energy demand	[kJ/s]
$\mathbb{E}[TH_i]$	Expected throughput of machine i	[ppm]
$\mathbb{E}[TH]$	Expected throughput of the production line	[ppm]

The following assumptions are assumed to be valid:

- The interarrival time is a random variable T_a with the probability density function (PDF) $f_a(t)$ modeling the time between two part arrivals at the station—i.e., the expected interarrival time is $\mathbb{E}[T_a] = \bar{t}_a$;
- The machine i processing time $T_{p,i}$ is random with the PDF $f_{p,i}(t)$—i.e., the expected processing time is $\mathbb{E}[T_{p,i}] = \bar{t}_{p,i}$;
- The start-up duration $T_{\text{su},i}$ of machine i is random with the PDF $f_{\text{su},i}(t)$—i.e., the expected start-up time is $\mathbb{E}[T_{\text{su},i}] = \bar{t}_{\text{su},i}$;
- Buffer i with finite capacity K_i controls the release of parts to the machine i and the queueing rule does not influence the developed analysis;
- There is an infinite capacity buffer downstream machine m, and thus the last machine cannot be blocked;

- The stochastic processes involved in the system are assumed to be independent of each other;
- Each machine is assumed to be perfectly reliable, thus failures cannot occur. This assumption can be relaxed without requiring large extensions to the developed analysis;
- All power and time values considered in the model are finite.

Let be $x_i \in \mathscr{S}_X = \{1, 2, 3, 4\}$ the i-th machine state: $x_i = 1$ if out-of-service, $x_i = 2$ if on-service, $x_i = 3$ if start-up, and $x_i = 4$ if working. Moreover, let $w_{x_i,i}$ be the power absorbed by the i-th machine in state x_i. In the *out-of-service state*—i.e., the standby state—the machine is in a kind of "sleeping" mode and it is not able to produce. The power demand of the machine when out-of-service is denoted with $w_{1,i}$, generally, lower compared to that in the other machine states. In the *on-service* state—i.e., the idle state—the machine is ready to process a part. The machine power demand when on-service, denoted with $w_{2,i}$, is due to the activation of all machine modules that have to be ready for processing a part. From the out-of-service to the on-service state, the machine must pass through the *start-up* state—i.e., the transitory state where a warming-up procedure is executed to make the modules suitable for processing. In this work, this transitory is referred as "start-up." The power demand $w_{3,i}$ is generally greater compared to that in the other machine nonproductive states. The start-up duration $T_{\mathrm{su},i}$ is assumed to be a random value because the machine may request variable time to reach the proper physical working condition according to the system status—e.g., room temperature. If the start-up is long enough to ensure machine proper conditions to work, the quality of processed parts is guaranteed. In the *working* state, the machine is processing a part and the requested power $w_{4,i}$ varies according to the process. If a part arrives when the machine is on-service, it can be immediately processed, otherwise it waits in the buffer with a power demand $w_{q,i}$ necessary for holding each part. The transition between two states can be triggered by the occurrence of an uncontrollable event—e.g., the part arrival—or a controllable event.

8.3 Control Policies

A general control policy, recently proposed in the literature (Frigerio and Matta 2014a), can be applied on each machine of the line. This general control policy is now presented describing the machine-buffer behavior in terms of states visited and transitions triggered.

Policy 1 (TNT-policy) If the buffer is empty, switch-off the machine after a time interval $\tau_{\mathrm{off},i}$ has elapsed from the last departure. Then, switch-on the machine when the number of parts in the queue reaches a proper level N_i or after a time interval $\tau_{\mathrm{on},i}$ has elapsed from the last departure—i.e., when $\tau_{\mathrm{on},i} - \tau_{\mathrm{off},i}$ has elapsed from the switch-off command.

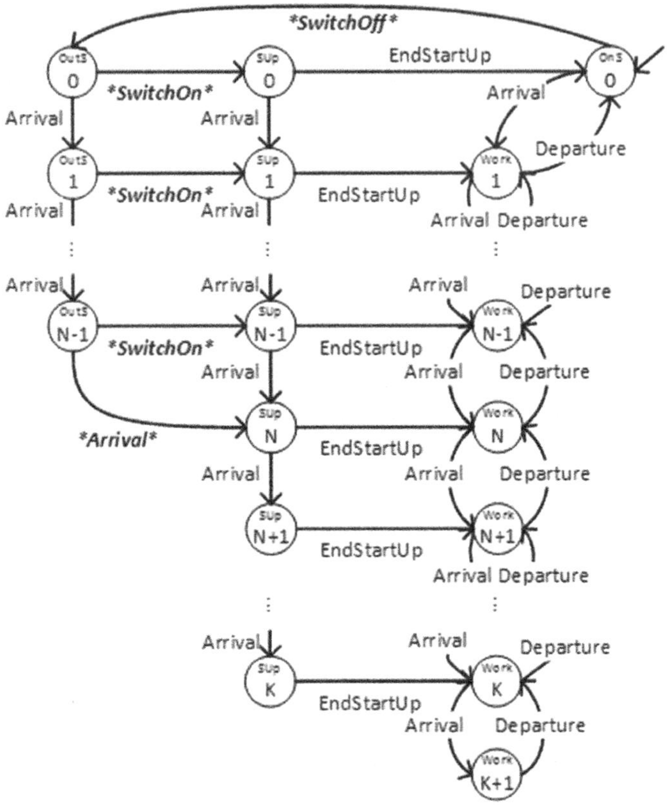

Fig. 8.3 Station event-based model: $x = 1$ out-of-service (OutS), $x = 2$ on-service (OnS), $x = 3$ start-up (SUp), and $x = 4$ working (Work)

The behavior of the station i is represented in Fig. 8.3. From the initial on-service state ($n_i = 0$), two situations may happen: an arrival occurs and the machine starts working the part, or the time interval $\tau_{\text{off},i}$ has elapsed and the machine is switched off. During the processing time, other parts can enter into the system until the station is full and the machine continues processing the parts in the queue until the last part is cleared from the system. The number n_i of parts in the system is incremented and decremented by arrivals and departures, respectively. Once in out-of-service, the machine is warmed up when the number of parts in the queue reaches a certain level $N_i \in [1, K_i]$. Otherwise, after $\tau_{\text{on},i}$ the machine can be switched on in advance to be ready to process the next part, even if the queue did not accumulate N_i parts. If the machine is in out-of-service, but not switched on yet, the system cannot contain more parts then $n_i = N_i - 1$, because the machine is switched on upon the N_i-th arrival. Moreover, in order to be consistent, the switch-off command has to be issued before the switch-on command as follows:

$$\tau_{on,i} \geq \tau_{off,i}. \tag{8.1}$$

If a part arrives while the machine is in the start-up state, the part must wait in the queue. The station cannot contain more parts then $n_i = K_i$ if the machine is executing the start-up because there are only K_i positions available for holding parts inside the buffer i. As the start-up procedure completes, the machine enters in the on-service state, if the station is empty ($n_i = 0$), or in a working state, otherwise.

Defining $t_{d,i}$ as the time instant of the last departure at station i, the TNT-policy at machine i is as follows.

$$\alpha_{TNT} = \begin{cases} \text{switch-off} & \text{if } x_i = 2 \wedge t - t_{d,i} = \tau_{off,i} \\ \text{switch-on} & \text{if } x_i = 1 \wedge (n_i = N_i \vee t - t_{d,i} = \tau_{on,i}) \\ \text{do nothing} & \text{otherwise.} \end{cases} \tag{8.2}$$

Moreover, $\mathbf{c}_i = \{\tau_{off,i}, N_i, \tau_{on,i}\}$ is defined as the vector composed by the three control parameters applied at machine i.

8.3.1 Special Cases

A *TNT-policy* is applied at machine level such that the machine is switched off to save energy, and, once in out-of-service, it is warmed up to resume the service properly. Some simpler strategies for managing a machine tool are discussed as special cases of Policy 1. The control parameters are summarized in Table 8.2.

The extreme situation in which the machine is always kept on-service is considered as the *Always on* policy (P2): The machine is never switched off neither the production is delayed. This control type represents all machines that do not have 'green' functionalities, or alternatively these situations where the controllable machine tool

Table 8.2 Control parameters for special case policies

Control policy	(P#)	Control parameter	Number of control parameters
TNT-policy	P1	$\mathbf{c} = \{\tau_{off}, N, \tau_{on}\}$	3
Always On	P2	$\mathbf{c} = \{\infty, \forall, \infty\}$	0
N-policy	P3	$\mathbf{c} = \{0, N, \infty\}$	1
TN-policy	P4	$\mathbf{c} = \{\tau_{off}, N, \infty\}$	2
NT-policy	P5	$\mathbf{c} = \{0, N, \tau_{on}\}$	2
Off	P6	$\mathbf{c} = \{0, 1, \infty\}$	0
Switch-off policy	P7	$\mathbf{c} = \{\tau_{off}, 1, \infty\}$	1
Switch-on policy	P8	$\mathbf{c} = \{0, 1, \tau_{on}\}$	1
Switching policy	P9	$\mathbf{c} = \{\tau_{off}, 1, \tau_{on}\}$	2

should not be switched off. The *N-policy* (P3) represents a policy where the machine stays out-of-service if there is no part to work. The machine is switched on when the station has accumulated a certain amount of work, therefore, the machine starts the warming-up procedure. Two hybrid policies are the *TN-policy* (P4) and the *NT-policy* (P5) where the control parameter N is coupled with τ_{off} or τ_{on}, respectively. Moreover, policies that do not consider buffer level information can also be considered as policies where $N = 1$. Indeed, the *Off* policy (P6) represents a special case of the *N-policy*. The *Switch-off* policy (P7), the *Switch-on* policy (P8), and the *Switching* policy (P9) are particular cases of a *TN-policy*, *NT-policy*, and *TNT-policy*, respectively.

8.3.2 Energy Measures

The machine expected energy demanded per part $\mathbb{E}[Q_i]$ is the ratio between the expected power and the expected throughput at the same workstation i:

$$\mathbb{E}[Q_i] = \frac{\mathbb{E}[P_i]}{\mathbb{E}[TH_i]} \left[\frac{kJ}{part}\right] \qquad i = 1, \ldots, m \tag{8.3}$$

The expected total energy demanded per part $\mathbb{E}[Q_{\text{tot}}]$ is the sum of the machine expected energy demands:

$$\mathbb{E}[Q_{\text{tot}}] = \sum_{i}^{m} \mathbb{E}[Q_i] \left[\frac{\text{kJ}}{\text{part}}\right] \tag{8.4}$$

In the following, the energy demand values are reported in kJ per part (kJ/part).

The workstation state is defined as $y_i = \{x_i, n_i\} \in \mathcal{S}_i$, where $\mathcal{S}_i$ is the irreducible set of feasible states of the workstation i as in Fig. 8.3. Therefore, the power demanded by the station i in state y_i can be defined as:

$$h(y_i) = \begin{cases} w_{x,i} + w_{q,i}(n_i - 1) & \text{if } x_i = 4 \\ w_{x,i} + w_{q,i} n_i & \text{if } x_i \neq 4 \end{cases} \tag{8.5}$$

As a consequence:

$$\mathbb{E}[P_i] = \sum_{y_i \in \mathcal{S}_i} Pr(Y_i = y_i) h(y_i) \, [kW] \qquad \forall i = 1, \ldots, m \tag{8.6}$$

$$\mathbb{E}[TH_i] = \frac{60}{\mathbb{E}[T_{p,i}]} \sum_{y_i \in \mathcal{S}_i \wedge x_i = 4} Pr(Y_i = y_i) \, [ppm] \qquad \forall i = 1, \ldots, m \tag{8.7}$$

The system throughput is defined as the throughput of machine m, $\mathbb{E}[TH_m] = \mathbb{E}[TH]$, and, due to the conservation of flow, it is $\mathbb{E}[TH_m] = \mathbb{E}[TH_i]$ $\forall$ $i = 1, \ldots, m-1$. In the following, the throughput values are reported in parts per minute (ppm).

8.4 Simulation Model

A template model—i.e., a library—has been created in ARENA© environment for modeling a general machine controlled by a general policy that considers the information from the machine input buffer. By using the template, the developer can rapidly build complex simulation models of production systems composed of energy-oriented controlled machines (Fig. 8.4) together with other elements and blocks usually available in the discrete event simulation software.

Each station is represented by a single block where the buffer and the controlled machine are embedded. The template is composed of around 60 embedded blocks and can be easily integrated with the rest of the simulation model using standard instructions. In more details, the template models the part flow through the buffer and the machine. The switch-off and switch-on commands are issued using signals.

The instantaneous buffer level and the current machine state can be directly visualized as in Fig. 8.5, together with the cumulative number of restarts executed from the simulation beginning and the average energy demanded per part. The machine parameters can be input using a convenient user interface in which the developer can introduce several data. In the first panel, labeled *"GreenMach"*, the user must give a

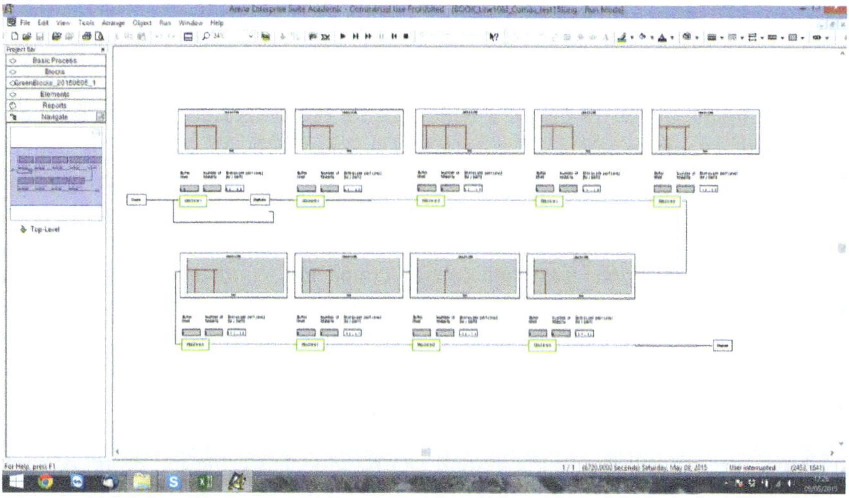

Fig. 8.4 Example of a production serial line built in ARENA©

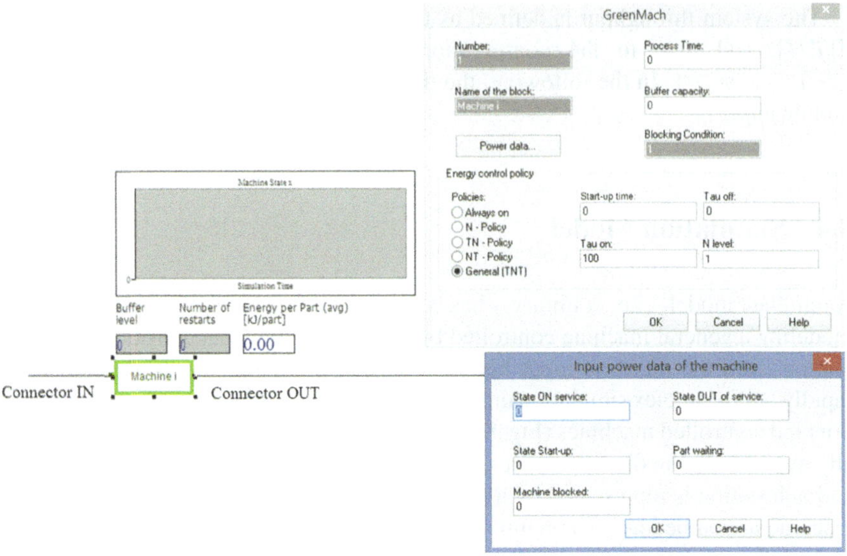

Fig. 8.5 "Green Machine" template with user interface and power input panel

unique name to the block as well as a sequential number: These elements characterize uniquely the block in the simulation model. The (finite) buffer capacity should also be inserted. Furthermore, the power demand in each machine state and the holding power demand can be set in the power data panel, labeled *"Input power data of the machine"*. The processing time can be modeled as a random variable. Specifically, the processing time distribution can be set according to ARENA© expressions for stochastic processes, e.g., the processing time input can be "EXPO($\bar{t}_p$)." The start-up time is also modeled as a random variable such that any distribution can be set as input.

The template allows to model a wide range of controls since five policies can be selected: *Always on* (P2), *N-Policy* (P3), *TN-Policy* (P4), *NT-Policy* (P5), and the general *TNT-policy* (P1). For each policy, the control parameters can be specified. All other policies described in Sect. 8.3—i.e., from P6 to P9—can be implemented as special cases with $N = 1$.

The *"Green Machine"* template can be connected to other simulation blocks through an input connector and an output connector. If the green machine cannot be blocked because the part can always be unloaded, the blocking condition is set to"`1`." Otherwise, the blocking condition should be specified. The power demand when the machine is blocked should also be set in the power panel. For example, it can be considered a system of two machines in series with a finite buffer each. Both machines are controlled and the system can be modeled using the proposed template. Assume that the template block `Machine(1)` is connected downstream with `Machine(2)`, and that `Machine(2)_K` is the buffer capacity of `Machine(2)`. The blocking condition on `Machine(1)` is:

"NQ(Machine(2)_buffer) <Machine(2)_K"

whereas the blocking condition on Machine(2) is "1".

8.5 Optimization of Production Line with Buffer Information

A production line composed of single machine workstations is the system to be analyzed in this work. The machine tools considered are CNC machining centers with small-medium workspace cube and locally chilled.

Two main control strategies at system level are considered in the experiments: the *single control* of the system, and the *complete control* of the system. As far as the single control, the system can be partially optimized by controlling the state of one machine, i.e., minimizing $\mathbb{E}[Q_i]$ at a certain workstation i. Considering one machine at the time, m optimizations are needed for each scenario. In the complete control, the system is globally optimized by controlling all machines simultaneously, i.e., by minimizing $\mathbb{E}[Q_{\text{tot}}]$. As a consequence, one optimization is needed for each scenario.

The goal is to assess the impact of controlling a specific machine instead of another, and the advantage of controlling the whole production line simultaneously. The starvation/blocking effect of the control will also be investigated. The *Always on* represents the reference case for evaluating the impact of the control.

8.5.1 Production Line Data

A real CNC machining center with 392 dm^3 of workspace, five axes, horizontal synchronous spindle, and local chiller—cooling both spindle and axes—is considered. This machine type requires $w_1 = 5.35$ kW while on-service and blocked, and $w_2 = 0.52$ kW when out-of-service. The machine start-up is characterized by a power demand of $w_3 = 6$ kW and a deterministic duration of $t_{\text{wu}} = 20$ s. These data have been acquired with dedicated experimental measurements from a real machine configured to operate in a powertrain manufacturing line. The power demanded during the working state is not considered in the analysis because it does not affect the selection of the control parameters, being not dependent on control actions. Each machine has the same power demand per stat—i.e., $w_{1,i} = w_1, w_{2,i} = w_2$, and $w_{3,i} = w_3$. Each buffer in the line has the same finite holding capacity and power demand $w_{q,i} = w_q$. Since all machines and buffers are identical, the production line is balanced with no bottleneck.

Two main scenarios are considered in the experiments: Case A, a line with three stations, and Case B, a line with nine stations. The buffer capacity is equal to ten slots at each workstation of the lines. In Case A, the part inter-arrivals at the first machine are assumed to be exponentially distributed with mean $\bar{t}_a = 110$ s. The processing time is assumed to be deterministic and equal to $t_p = 100$ s. Furthermore, to evaluate how the control parameters are affected by the holding power demand at buffers, the experiments are repeated for $w_q = 0$ kW and $w_q = 0.1$ kW. In Case B, an holding power demand $w_q = 0.1\ kW$ is considered and it is assumed that the first machine is never starved for raw material. As a consequence, the first machine (M1) is kept *Always on*. Therefore, the energy demanded by M1 is only associated with the blocking condition and the holding power demand. The processing times are identically distributed and follow a discrete distribution with two values: 100 s in 95% of the cases and 280 s in 5%. With this distribution, it is modeled in a simplified way a server with failure durations of the length of 3 min in addition to the standard processing times of 100 s.

8.5.2 Numerical Experiments

The simulation model of the analyzed production lines was developed in ARENA©. In these experiments, the control parameters are optimized to find out the optimal values that minimize the energy demand. OptQuest is used in order to select a set of nearly optimal solutions by varying the three control parameters (τ_{off}, N, and τ_{on}) for the controlled machines. OptQuest automatically compares solutions to be significantly different on average (95% of confidence), and it selects the best solution. However, some solutions can be statistically equivalent according to the simulation results. Moreover, the interval of confidence is verified to be less than 5% around the mean during the optimizations. Simulations were executed for 115 days with an initial transient identified with the Welch method (6 days) as detailed in Table 8.3. The solutions identified by OptQuest have been evaluated by running five independent replications, and the results are reported in Tables 8.4, 8.5, 8.6, 8.7, 8.8, 8.9, and 8.10. The optimization duration is also reported, running on a notebook DELL XPS15 with Intel Core @2.20 GHz and 8GB RAM.

Table 8.3 Optimization duration for each case

Case	A-Single	A-Complete	B-Single	B-Complete
Start-up length	6 days	6 days	6 days	6 days
Simulation length	115 days	115 days	115 days	115 days
Policy optimized	P1	P1	P1	P1
Max # iteration	50	100	100	1000 (150)
Optimization duration	6 min	15 min	60 min	8h (1h)

Table 8.4 Energy and throughput Case A—Always on

w_q (kW)	$\mathbb{E}[Q_1]$	$\mathbb{E}[Q_2]$	$\mathbb{E}[Q_3]$	$\mathbb{E}[Q_{tot}]$	$\mathbb{E}[TH]$
0	61.903 ± 3.310	61.891 ± 3.309	61.896 ± 3.315	185.681 ± 9.935	0.5376 ± 0.0011
0.1	94.983 ± 2.672	61.891 ± 3.309	61.896 ± 3.315	218.227 ± 9.269	0.5376 ± 0.0011

Table 8.5 Energy and throughput Case A ($w_q = 0\,kW$)—Controlled

	Single control			Complete control	Combined control
	on M1	on M2	on M3		
	$c_1^* = \{0, 6, \infty\}$	$c_1 = \{\infty, *, \infty\}$	$c_1 = \{\infty, *, \infty\}$	$\mathbf{c}^* = (\{0, 3, \infty\}; \{0, 5, \infty\}; \{0, 10, \infty\})$	$\mathbf{c}^* = (\{0, 6, \infty\}; \{0, 10, \infty\}; \{0, 10, \infty\})$
	$c_2 = \{\infty, *, \infty\}$	$c_2^* = \{0, 10, \infty\}$	$c_2 = \{\infty, *, \infty\}$		
	$c_3 = \{\infty, *, \infty\}$	$c_3 = \{\infty, *, \infty\}$	$c_3^* = \{0, 10, \infty\}$		
$\mathbb{E}[Q_1]$	8.996 ± 3.373	61.893 ± 3.310	61.903 ± 3.310	10.067 ± 0.493	8.996 ± 0.373
$\mathbb{E}[Q_2]$	70.611 ± 3.022	7.013 ± 0.382	61.891 ± 3.309	8.208 ± 0.411	7.936 ± 0.305
$\mathbb{E}[Q_3]$	70.611 ± 3.315	61.877 ± 3.022	7.013 ± 0.382	7.268 ± 0.355	7.936 ± 0.352
$\mathbb{E}[Q_{tot}]$	150.218 ± 6.41	130.784 ± 7.034	130.798 ± 7.034	25.454 ± 1.259	24.869 ± 1.077
	−19.29%	−29.56%	−29.55%	−86.26%	−86.61%
$\mathbb{E}[TH]$	0.5299 ± 0.0010	0.5376 ± 0.0011	0.5376 ± 0.0011	0.5351 ± 0.0010	0.5294 ± 0.0010
	−1.43%	–	–	−0.45%	−1.52%

8.6 Numerical Results

8.6.1 Case A: A Line with 3 Machines

8.6.1.1 Always on Analysis

All machines are equally utilized and there is no bottleneck. The energy demanded by each machine is around 62.1 kJ/part on average and the throughput is around 0.54 ppm as reported in Table 8.4. Parts accumulate in front of the first machine because of arrival variability, whereas there is no queue in front of M2 and M3 because theline

Table 8.6 Energy and throughput Case A ($w_q = 0.1\ kW$)—Controlled

	Single control			Complete control	Combined control
	on M1	on M2	on M3		
	$c_1^* = \{0, 2, \infty\}$	$c_1 = \{\infty, *, \infty\}$	$c_1 = \{\infty, *, \infty\}$	$\mathbf{c}^* = (\{0, 3, \infty\}; \{0, 1, \infty\}; \{0, 1, \infty\})$	$\mathbf{c} = (\{0, 2, \infty\}; \{0, 1, \infty\}; \{0, 1, \infty\})$
	$c_2 = \{\infty, *, \infty\}$	$c_2^* = \{0, 1, \infty\}$	$c_2 = \{\infty, *, \infty\}$		
	$c_3 = \{\infty, *, \infty\}$	$c_3 = \{\infty, *, \infty\}$	$c_3^* = \{0, 1, \infty\}$		
$\mathbb{E}[Q_1]$	52.211 ± 0.416	94.438 ± 2.672	94.438 ± 2.152	49.625 ± 0.412	52.211 ± 0.416
$\mathbb{E}[Q_2]$	64.656 ± 3.201	17.643 ± 0.828	61.891 ± 3.309	13.552 ± 0.546	12.068 ± 0.493
$\mathbb{E}[Q_3]$	64.656 ± 3.201	61.896 ± 3.315	17.644 ± 0.829	13.552 ± 0.546	12.068 ± 0.493
$\mathbb{E}[Q_{\text{tot}}]$	181.525 ± 6.452	173.979 ± 6.794	173.975 ± 6.788	76.731 ± 1.098	76.347 ± 1.088
	−16.81%	−20.28%	−20.28%	−64.83%	−65.01%
$\mathbb{E}[TH]$	0.5363 ± 0.0010	0.5376 ± 0.0011	0.5376 ± 0.0011	0.5363 ± 0.0011	0.5351 ± 0.0011
	−0.24%	–	–	−0.24%	−0.7%

Table 8.7 Average energy demand for Case B—Always on

Station #	On-Service En.	Holding En.	Blocking En.	Total energy
$\mathbb{E}[Q_1]$	0	112.050	15.244	127.296 ± 1.450
$\mathbb{E}[Q_2]$	3.122	73.942	11.812	88.877 ± 2.620
$\mathbb{E}[Q_3]$	4.641	66.990	10.289	81.920 ± 3.790
$\mathbb{E}[Q_4]$	6.070	61.639	9.217	76.928 ± 3.711
$\mathbb{E}[Q_5]$	6.987	58.987	8.373	73.985 ± 2.786
$\mathbb{E}[Q_6]$	8.052	55.847	6.373	70.272 ± 1.953
$\mathbb{E}[Q_7]$	10.157	50.011	5.089	65.257 ± 1.774
$\mathbb{E}[Q_8]$	11.523	46.490	3.263	61.277 ± 0.876
$\mathbb{E}[Q_9]$	14.696	39.548	0	54.244 ± 1.696
$\mathbb{E}[Q_{\text{tot}}]$	65.248	563.912	67.534	700.056 ± 15.566

is synchronous and the second and third buffers are empty. Indeed, considering a holding demand of $w_q = 0.1$ kW per part, the first station consumes more than other stations (94.983 kJ/part), whereas the energy demand of M2 and M3 is unvaried.

Table 8.8 Average energy demand Case B—Single control ($w_q = 0.1$ kW)

Control on	$\mathbb{E}[Q_i]$	$\mathbb{E}[Q_{\text{tot}}]$	$\mathbb{E}[TH]$
M2 $c_2^* = \{0, 1, \infty\}$	87.687(±1.143) −1.34%	690.902(±7.876) −1.31%	0.5368(±0.0006) −0.001%
M3 $c_3^* = \{0, 1, \infty\}$	76.049(±3.031) −7.17%	686.300(±13.075) −1.96%	0.5370(±0.0005) −0.004%
M4 $c_4^* = \{0, 1, \infty\}$	73.644(±0.123) −4.27%	692.314(± 10.231) −1.11%	0.5372(±0.0005) −0.04%
M5 $c_5^* = \{0, 1, \infty\}$	67.980(±5.217) −8.12%	682.756(±12.458) −2.47%	0.5372(±0.0004) −0.04%
M6 $c_6^* = \{0, 1, \infty\}$	60.767(±3.022) −13.53%	678.801(±16.334) −3.04%	0.5370(±0.0004) −0.004%
M7 $c_7^* = \{0, 1, \infty\}$	57.684(±4.164) −12.29%	690.826(±22.092) −1.31%	0.5370(±0.0003) −0.004%
M8 $c_8^* = \{0, 1, \infty\}$	52.684(±1.201) −14.02%	685.105(±16.108) −1.99%	0.5371(±0.0001) −0.02%
M9 $c_9^* = \{0, 1, \infty\}$	44.350(±1.257) −18.24%	695.988(±6.396) −0.58%	0.5370(±0.0004) −0.004%

Table 8.9 Energy and throughput Case B—Optimal complete control ($w_q = 0.1kW$)

	Average	IC	Δ% (%)	c_i
$\mathbb{E}[Q_1]$	130.854	±1.035	2.80	{∞; ∗; ∞}
$\mathbb{E}[Q_2]$	99.089	±2.586	11.49	{0; 9; ∞}
$\mathbb{E}[Q_3]$	88.047	±2.572	7.48	{0; 8; ∞}
$\mathbb{E}[Q_4]$	60.815	±6.176	−20.95	{0; 1; ∞}
$\mathbb{E}[Q_5]$	54.878	±5.832	−25.83	{0; 1; ∞}
$\mathbb{E}[Q_6]$	53.359	±4.669	−24.07	{0; 1; ∞}
$\mathbb{E}[Q_7]$	50.257	±4.294	−22.99	{0; 1; ∞}
$\mathbb{E}[Q_8]$	45.682	±2.235	−25.45	{0; 1; ∞}
$\mathbb{E}[Q_9]$	39.778	±0.836	−26.67	{0; 1; ∞}
$\mathbb{E}[Q_{\text{tot}}]$	622.758	±22.573	−11.04	
$\mathbb{E}[TH]$	0.5341	±0.0006	−0.5	

8.6.1.2 Single Control ($w_q = 0$ kW)

The single control is applied at each workstation one-by-one varying the machine control parameters. Compared to the standard *Always on* policy, a control policy may decrease the utilization of the controlled machine because of the period spent in the out-of-service state. However, the impact on the rest of the system is not easy to evaluate. For these reasons, during simulation, the energy demands are collected along the line, as well as the line throughput (see Table 8.5). The optimal control parameters for each machine tool are also reported.

Table 8.10 Energy and throughput Case B—Constrained optimal complete control ($w_q = 0.1$ kW)

	Average	IC	Δ% (%)	c_i
$\mathbb{E}[Q_1]$	127.054	±1.580	−0.19	$\{\infty; *; \infty\}$
$\mathbb{E}[Q_2]$	89.968	±3.368	1.23	$\{0; 5; \infty\}$
$\mathbb{E}[Q_3]$	73.068	±3.650	−10.81	$\{0; 1; \infty\}$
$\mathbb{E}[Q_4]$	68.368	±3.560	−11.13	$\{0; 1; \infty\}$
$\mathbb{E}[Q_5]$	65.505	±1.737	−11.46	$\{0; 1; \infty\}$
$\mathbb{E}[Q_6]$	62.172	±2.730	−11.53	$\{0; 1; \infty\}$
$\mathbb{E}[Q_7]$	57.760	±3.553	−11.49	$\{0; 1; \infty\}$
$\mathbb{E}[Q_8]$	52.081	±2.787	−15.01	$\{0; 1; \infty\}$
$\mathbb{E}[Q_9]$	44.023	±1.186	−18.84	$\{0; 1; \infty\}$
$\mathbb{E}[Q_{tot}]$	640.000	±15.000	−8.58	
$\mathbb{E}[TH]$	0.5370	±0.0003	−0.02	

When the single control is optimized, the *TNT-policy* degenerates in an *N-policy*. For the first machine (M1), the minimum energy demanded is associated with a control $c_1^* = \{0, 6, \infty\}$. It is noteworthy that a state control on a machine may influence the energy demanded by the downstream machines ($\mathbb{E}[Q_2]$ and $\mathbb{E}[Q_3]$) which are starving due to the reduced throughput. As a consequence, the downstream machines, kept on-service longer, increase their energy demand.

For the second machine (M2), the optimal control is $c_2^* = \{0, 10, \infty\}$; hence, M2 is kept out-of-service longer compared to M1. Generally, the upstream machine is subject to higher blocking probability because the number of parts in the queue of the controlled machine increases. However, in this case, this effect is negligible. It is similarly for the control on M3 ($c_3^* = \{0, 10, \infty\}$).

A comparison with the *Always on* scenario shows that the single control achieves significant energy saving without significantly affecting the line throughput (Kruskal–Wallis test with $p_{\text{value}} < 0.05$). The Mann-Whitney test confirms that any single control reduces significantly the energy demanded at system level ($p < 0.05$). Moreover, the position of the controlled machine is also significant: Controlling M2 has the same effect as controlling M3, whereas the control on M1 performs better (Mann-Whitney $p_{\text{value}} < 0.05$) in this case.

8.6.1.3 Complete Control ($w_q = 0$ kW)

When all machines are controlled simultaneously, as reported in Table 8.5, the optimal control $\mathbf{c}^* = (c_1; c_2; c_3) = (\{0, 3, \infty\}; \{0, 5, \infty\}; \{0, 10, \infty\})$ switches off/on all machines with an *N-Policy*.

The complete control performs better than any single control achieving around 87% of average energy reduction (Mann-Whitney test with $p_{\text{value}} < 0.05$). However, the combined control obtained by locally optimizing each machine $\mathbf{c} = (c_1^*; c_2^*; c_3^*) = (\{0, 6, \infty\}; \{0, 10, \infty\}; \{0, 10, \infty\})$ is not significantly different from the optimal solution $\mathbf{c}^* = (\{0, 3, \infty\}; \{0, 5, \infty\}; \{0, 10, \infty\})$ identified by OptQuest.

8.6.1.4 Holding Power Effect

Now, it is assumed that the holding power of buffers in the line is $w_q = 0.1\ kW$ (see Table 8.6). As a consequence, the energy demanded for holding parts in the buffers affects the optimal control parameters since it is increasing in N. Generally, the optimal control policy is an *N-Policy*, but the service is resumed sooner compared to the previous optimization—i.e., N decreases compared to Table 8.5—to avoid the holding power demand. The Kruskal–Wallis test confirms that the single control reduces the energy demanded at system level ($p_{\text{value}} < 0.05$). Moreover, the position of the controlled machine—i.e., to control M1 instead of others—is also significant. The complete control performs better then any single control (Mann-Whitney test with $p_{\text{value}} < 0.05$). However, the combined control obtained by locally optimizing each machine $\mathbf{c} = (c_1^*; c_2^*; c_3^*) = (\{0, 2, \infty\}; \{0, 1, \infty\}; \{0, 1, \infty\})$ is not significantly different from the optimal solution $\mathbf{c}^* = (\{0, 3, \infty\}; \{0, 1, \infty\}; \{0, 1, \infty\})$ identified by OptQuest.

8.6.2 *Case B: A Line with 9 Machines*

8.6.2.1 Always on Analysis

As described in Sect. 8.5.1, the first machine is always fed: Its demand is related only to blocking periods. The machine utilization decreases along the line as well as the average queue length and the blocking probability. In Table 8.7, the energy demanded by each machine while on-service and while blocked, and by the parts waiting in the buffer are reported. For this scenario, the total energy demand trend is decreasing because the most important component is the holding power demand: the far the machine in the line (higher i), the lower its energy demand. It is noteworthy that the energies demanded in the blocking state and in the on-service state are symmetric because the machine power demand is the same during starvation and blocking, and because starvation and blocking effects are symmetric when the machines are not controlled. The total average energy per part demanded by the whole line is around 700 kJ/part and the production rate is 0.5368 ± 0.0010 ppm.

8.6.2.2 Single Control

Similarly to Case A, the system performance is collected in Table 8.8. The optimal control c_i^* minimizing the machine energy $\mathbb{E}[Q_i]$ is also reported for each optimization. The single machine control influences the energy demanded by other stations in the line. Thus, two effects can be identified: (i) starvation effect downstream and (ii) blocking effect upstream.

(i) The downstream machines are less utilized because of the reduced throughput of the controlled machine. As a consequence, the holding energy is reduced because buffers are empty, the energy demanded by the machines on-service increases, and the energy in blocked condition decreases.
(ii) The number of parts in the queue of the controlled machine increases: The upstream machines have a higher blocking probability and longer queues.

As a consequence, the single control may achieve significant results on the controlled machine but not at line level. In order to give an example, the *Off* policy on M9 significantly decreases M9 energy demand (Mann-Whitney $p_{\text{value}} < 0.05$), while it is not effective on energy demand at line level (Mann-Whitney $p_{\text{value}} > 0.05$).

8.6.2.3 Numerical Fact

Before the optimization of the complete control strategy, an interesting numerical fact related to the blocking/starvation effect is discussed. It is assumed that the same control is applied to M2 to M9 except that the control parameter N_i is increased on the machine i. In order to give an example, Fig. 8.6 represents the machine energy demand along the line (M1 to M9). The control strategies represented are: the *Always on* case (black line/triangle), the case where the *N-policy* with $c = \{0, 1, \infty\}$ is applied to all machines M2–M9 (blue line/circle), and the case where the control of one machine j has a greater threshold compared to that of other machines $N_j \geq N_i\ \forall i \neq j$ (red line/cross). Particularly, the control $c_5 = \{0, 10, \infty\}$

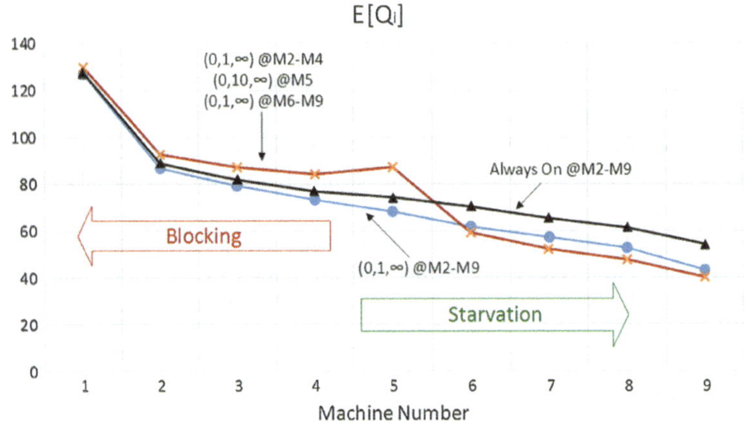

Fig. 8.6 Example of a production serial line built in ARENA©

has been applied to M5 whereas $c_i = \{0, 1, \infty\}$ has been applied to other machines ($i = 2, 3, 4, 6, 7, 8, 9$).

(i) The upstream machines (M1–M4) have a higher blocking probability and they cannot be ed off while blocked (as an assumption). Therefore, their energy demand increases.
(ii) The downstream machines (M6–M9) can be switched off because of starvation effect. Indeed, with the same control parameters, as far as the utilization decreases, a machine stay out-of-service longer and the average queue length decreases too. Nevertheless, the number of start-up increases, thus the energy demanded by the machines downstream the controlled one may decrease or increase according to this trade-off.

According to the numerical fact, increasing N at the last machines of the line negatively affects the energy demand upstream. Thus, for the case study under consideration, an optimal policy may have the following property:

$$N_i \geq N_j \qquad \forall i \leq j; \quad i, j = 1, \ldots, m \tag{8.8}$$

A complete control with this structure keeps the first machines in the line accumulating parts and doing less start-up procedures, whereas the last machines in the line start processing parts before to avoid blocking upstream.

8.6.2.4 Complete Control

The system performance under the optimal complete control is collected in Table 8.9. Although some machines increase their energy demand (M1–M3), this control achieves 11.04% of energy savings for the whole line, and the throughput is reduced by 0.5%. Particularly, M2 and M3 are controlled with an *N-policy* where $c_2 = \{0, 9, \infty\}$ and $c_3 = \{0, 8, \infty\}$, respectively. M4–M9 are controlled with an *Off* policy ($c = \{0, 1, \infty\}$). Also in this case, the control is significant according to the Mann-Whitney test ($p_{\text{value}} < 0.05$) as well as the reduction of system throughput. In this case, the combined control obtained by locally optimizing each machine with a *Off* policy (650.566 $\pm$ 12.129 kJ/part) is significantly different from the optimal solution (622.758 $\pm$ 22.573 kJ/part).

It is noteworthy that OptQuest needed around 1000 iterations to identify the best solution (see Table 8.3). However, it identified another solution that is not significantly different (Mann-Whitney test $p_{\text{value}} = 0.60$) after 150 simulations (1h). This solution is an *N-policy* where $c_2 = \{0, 10, \infty\}$ and $c_3 = \{0, 8, \infty\}$, for M2 and M3 respectively, and M4–M9 are controlled with an *Off* policy ($c = \{0, 1, \infty\}$).

When a constraint for the minimum target throughput is included in the optimization model, the optimal policy does not change structure. For instance, with a minimum throughput of 0.5364 ppm (0.1% of reduction compared to the *Always on*), the constrained optimal control achieves 8.58% of savings on the total energy demanded. Actually, in this case, the control does not significantly affect the production rate (Mann-Whitney $p_{\text{value}} = 0.40$). Particularly, M3 to M9 are controlled with an *N-policy* $c^* = \{0; 1; \infty\}$ and M2 with $c_2^* = \{0; 5; \infty\}$, as in Table 8.10. Compared to the Case A, the energy savings are lower because of the high variability of the machines. Indeed, the variable processing times give place to the presence of parts in the intermediate buffers that do not allow a frequent switch-off of machines.

8.7 Unbalanced Lines

An unbalanced line (Case C and Case D) has been studied when a complete control is applied. Cases C and D are the same as Case B except that one machine can be identified as bottleneck. Indeed, one machine has a different processing time: 100 s in

Table 8.11 Average energy demand for Case C—Always on

Station #	On-Service En.	Holding En.	Blocking En.	Total energy
$\mathbb{E}[Q_1]$	0	118.979	48.607	167.780 ± 1.504
$\mathbb{E}[Q_2]$	0.282	103.359	48.875	152.390 ± 1.889
$\mathbb{E}[Q_3]$	0.284	103.311	48.860	152.350 ± 0.977
$\mathbb{E}[Q_4]$	0.378	103.355	48.307	152.182 ± 2.514
$\mathbb{E}[Q_5]$	0.351	102.295	48.636	152.642 ± 2.426
$\mathbb{E}[Q_6]$	0.216	103.314	48.813	151.704 ± 3.084
$\mathbb{E}[Q_7]$	0.246	103.679	48.317	151.786 ± 2.712
$\mathbb{E}[Q_8]$	0.336	102.284	48.153	151.232 ± 2.968
$\mathbb{E}[Q_9]$	0.195	102.108	0	102.346 ± 1.285
$\mathbb{E}[Q_{\text{tot}}]$	2.289	943.981	388.565	1333.412 ± 17.032

Table 8.12 Average energy demand for Case D—Always on

Station #	On-Service En.	Holding En.	Blocking En.	Total energy
$\mathbb{E}[Q_1]$	0	119.609	48.607	167.826 ± 1.300
$\mathbb{E}[Q_2]$	0.258	103.790	48.032	152.702 ± 1.357
$\mathbb{E}[Q_3]$	0.391	103.546	49.123	152.574 ± 1.425
$\mathbb{E}[Q_4]$	0.353	103.471	48.939	152.092 ± 1.890
$\mathbb{E}[Q_5]$	0.158	102.761	0.193	102.850 ± 0.970
$\mathbb{E}[Q_6]$	49.082	15.180	0.263	64.372 ± 0.766
$\mathbb{E}[Q_7]$	48.959	15.862	0.278	64.908 ± 0.586
$\mathbb{E}[Q_8]$	49.302	15.750	0.202	64.921 ± 1.034
$\mathbb{E}[Q_9]$	48.618	15.994	0	64.467 ± 0.643
$\mathbb{E}[Q_{\text{tot}}]$	197.121	595.963	196.635	986.712 ± 8.471

Table 8.13 Energy and throughput Case D—Optimal complete control

	Average	IC	Δ% (%)	c_i
$\mathbb{E}[Q_1]$	165.912	±1.337	−1.14	$\{\infty; *; \infty\}$
$\mathbb{E}[Q_2]$	149.650	±1.358	−2.00	$\{\infty; *; \infty\}$
$\mathbb{E}[Q_3]$	148.858	±2.024	−2.44	$\{\infty; *; \infty\}$
$\mathbb{E}[Q_4]$	150.086	±1.237	−1.32	$\{\infty; *; \infty\}$
$\mathbb{E}[Q_5]$	102.786	±0.546	−0.06	$\{\infty; *; \infty\}$
$\mathbb{E}[Q_6]$	42.968	±0.482	−33.25	$\{0; 1; \infty\}$
$\mathbb{E}[Q_7]$	41.809	±0.788	−35.59	$\{0; 1; \infty\}$
$\mathbb{E}[Q_8]$	41.391	±0.331	−36.24	$\{0; 1; \infty\}$
$\mathbb{E}[Q_9]$	40.919	±0.676	−26.53	$\{0; 1; \infty\}$
$\mathbb{E}[Q_{\text{tot}}]$	884.380	±5.288	−10.37	
$\mathbb{E}[TH]$	0.5089	±0.0006	−0.19	

90% of the cases and 280 s in 10%. With this distribution, the stand-alone utilization of the machine is reduced of around 8%. In Case C, the bottleneck is the last machine (M9), whereas in Case D it is the central machine (M5). The OptQuest optimization results for the complete control of the scenarios C and D are collected in Tables 8.11, 8.12 and 8.13, respectively.

8.7.1 Case C

In this case, the bottleneck is M9 and all upstream machines (M1-M8) are frequently blocked and the average queue length is constant, compared to Case B where it is decreasing along the line (as well as blocking probability). Therefore, the machine energy demand is constant and mostly depends on the holding and the blocking energy requests (Table 8.11). The first machine (M1) is never starved and the queue is always full; therefore, it has a slightly higher energy demand. The last machine (M9) is never blocked and its energy is the lowest. The total average energy per part demanded by the whole line is 1333.412 ± 13.717 kJ/part and the production rate is 0.5082 ± 0.0010 ppm. The energy is greater than that of Case B because of the holding power demand, whereas the throughput is reduced because of the bottleneck. The optimal result by OptQuest (after 100 iterations—40 min optimization) is to apply the *Always on* control strategy (P2) to all machines, in order to avoid the starvation of the bottleneck.

8.7.2 Case D

In this case, machines upstream the bottleneck (M1–M4) are frequently blocked and the bottleneck (M5) is saturated, whereas machines downstream (M6–M9) are less utilized compared to Case B and they are often on-service. The first group of machines and the bottleneck have also a high holding energy demand (see Table 8.12). The total average energy per part demanded by the whole line is $986.712 \pm 8.471\ kJ/part$ and the production rate is the same as in Case C ($0.5081 \pm 0.0009\ ppm$). The optimal result by OptQuest (after 100 runs - 40 min optimization) is to apply the *Always on* control strategy (P2) to machines (M1–M5), in order to avoid the starvation of the bottleneck machine. Machines (M6–M9) are controlled with an *N-Policy*, as in Table 8.13, achieving around 35% of energy saving at each machine. The total average energy per part demanded by the whole line is reduced to $884.380 \pm 5.288\ kJ/part$ (-10.37%) without significantly affecting the productivity.

8.8 Remarks and Outlook

A general control policy with three parameters has been applied to machines in a production line with finite buffer capacities. The template built in ARENA© software environment allows to study complex systems when the machines are controlled. Comparing the controlled line performance with the common practice in manufacturing, it is remarkable that:

- A general policy may achieve significant energy savings without largely compromising the production throughput;
- To locally control each machine is sub-optimal, indeed, to control simultaneously the machines achieves the best results;
- The control of one machine influences the energy demanded by the other machines due to blocking and starvation effects;
- A control policy that is not uniform along the line may achieve significant results both for balanced line and line with bottleneck;
- In a balanced production line where each machine is controlled, the optimal parameter values of the threshold N may have a special structure such as decreasing;
- Machine variability affects the potential energy savings.

Since these results were obtained with simulation, there is no claim they are general and valid for any production line. Future developments will be devoted to generalize these results for other production systems. A sensitivity analysis for critical parameters is considered as an important extension for policy applicability. Furthermore, multiple low-energy states can be considered, and a multiple-standby energy control policy can be studied taking into account different start-up times for each machine transitory.

References

Abele E, Schrems S, Eisele C, Schraml P (2012) Simulation based assessment of the energy consumption of manufacturing processes. In: Proceedings of the 19th CIRP international conference on LCE, pp 375–379

Brundage MP, Chang Q, Li Y, Xiao G, Arinez J (2014) Energy efficiency management of an integrated serial production line and HVAC system. IEEE Trans Autom Sci Eng 12(3):789–797

Chang Q, Xiao G, Biller S, Li L (2013) Energy saving opportunity analysis of automotive serial production systems. IEEE Trans Autom Sci Eng 10(2):334–342

Chen G, Zhang L, Arinez J, Biller S (2011) Feedback control of machine startup for energy-efficient manufacturing in Bernoulli serial line. In: Proceedings of the IEEE conference on automation science and engineering, pp 666–671, Aug 2011

Chen G, Zhang L, Arinez J, Biller S (2013) Energy-efficient production systems through schedule-based optimization. IEEE Trans Autom Sci Eng 10(1):27–37

Dahmus JB, Gutowski TG (2010) An environmental analysis of machining. In: Manufacturing engineering and materials handling engineering (ASME), pp 643–652

DMG Energysave. http://nl.dmg.com/. Available via DIALOG

Frigerio N, Matta A (2014a) Energy saving policies for a machine tool with warm-up, stochastic arrivals and buffer information. In: IEEE international conference on automation science and engineering (CASE), pp 646–651, Aug 2014a

Frigerio N, Matta A (2014b) Energy efficient control strategy for machine tools with stochastic arrivals and time dependent warm-up. In: Proceedings of the 21th CIRP international conference on LCE, pp 56–61, June 2014b

Frigerio N, Matta A (2015) Energy efficient control strategies for machine tools with stochastic arrivals. IEEE Trans Autom Sci Eng **12**(1)

Ghadimi P, Li W, Kara S, Hermann C (2014) Integrated material and energy flow analysis towards energy efficient manufacturing. In: Proceedings of the 21th CIRP international conference on life cycle engineering (LCE), pp 117–122, June 2014

Guo X, Zhou S, Niu N, Kumar PR (2013) Optimal wake-up mechanism for single base station with sleep mode. In: 25th international teletraffic congress (ITC), pp 1–8, Sept 2013

Gutowski TG, Branham MS, Dahmus JB, Jones AJ, Thiriez A, Sekulic DP (2009) Thermodynamic analysis of resources used in manufacturing processes. Environ Sci Technol 43(5):1584–1590

Heidenhain (2011a) Aspects of energy efficiency in machine tool. http://www.heidenhain.co.uk/. Available via DIALOG

Heidenhain (2011b) Energy-efficient machine tools are economical and productive with Sinumerik Ctrl-Energy from Siemens. http://www.siemens.com/. Available via DIALOG

Heinemann T, Schraml P, Thiede S, Eisele C, Herrmann C, Abele E (2014) Hierarchical evaluation of environmental impact from manufacturing system and machine perspective. In: Proceedings of the 21th CIRP international conference on LCE, pp 141–146, June 2014

Ke JC, Wu CH, Zhang ZG (2010) Recent developments in vacation queueing models: a short survey. Int J Oper Res 7(10):3–8

Langer T, Schlegel A, Stoldt J, Putz M (2014) A model-based approach to energy-saving manufacturing control strategies. In: Proceedings of the 21th CIRP international conference on LCE, pp 123–128, June 2014

Maccio VJ, Down DG (2013) On optimal policies for energy-aware servers. In: IEEE 21st international symposium on modeling, analysis and simulation of computer and telecommunication systems (MASCOTS), pp 31–39, Aug 2013

Mashaei M, Lennartson B (2013) Energy reduction in a pallet-constrained flow shop through on-off control of idle machines. IEEE Trans Autom Sci Eng 10(1):45–56

Mouzon G, Yildirim MB, Twomey J (2007) Operational methods for minimization of energy consumption of manufacturing equipment. Int J Prod Res 45(18–19):4247–4271

Sun Z, Li L (2013) Opportunity estimation for real-time energy control of sustainable manufacturing systems. IEEE Trans Autom Sci Eng 10(1):38–44

Tian N, Zhang ZG (2006) Vacation queueing models—theory and applications. Springer
Weinert N, Mose C (2014) Investigation of advanced energy saving stand by strategies for production systems. In: Proceedings of the 21th CIRP international conference on LCE, pp 90–95, June 2014
Wu K, McGinnis L (2013) Interpolation approximations for queues in series. IIE Trans 45(3):273–290
Yadin M, Naor P (1963) Queueing systems with removable service station. Oper Res Q 14(4):393–405

Chapter 9
Approach for Achieving Transparency in the Use of Compressed Air in Manufacturing as a Basis for Systematic Energy Saving

Nico Pastewski, Susanne Krichel, Gerrit Posselt, Johannes Linzbach and Martin Plank

Abstract The increase in energy efficiency in manufacturing and in particular in the area of compressed air is becoming a top issue in the context of factory planning and operation. Facing this development, the European Commission entrusted the Seventh Framework Program, Horizon 2020, to strengthen the research efforts to hit the proclaimed environmental, energetic and economic objectives of the European Union. In this paper, the authors present selected results on how to re-engineer brownfield factories to become clean and competitive factories of the future. Furthermore, the focus is put on energy-efficient generation, preparation, distribution and utilization of compressed air. It provides a systematic and practical approach to re-engineer existing compressed air systems to become eco-efficient. In European industries, around 10% of the final energy utilized is converted into compressed air as a factory internal energy source and hence plays a vital role for various applications such as handling and assembly of parts. When generating compressed air, it is important to use efficient compressor stations with optimized compressed air preparation and appropriately designed compressed air systems for distributing and utilizing the compressed air. However, due to a lack of transparency and knowledge in terms of actual demands and technical dependencies, compressed air systems are

N. Pastewski (✉) · S. Krichel · J. Linzbach · M. Plank
Festo AG & Co. KG, Ruiter Strasse 82, 73734 Esslingen, Germany
e-mail: susanne.krichel@festo.com

G. Posselt
Chair of Sustainable Manufacturing and Life Cycle Engineering, Institute of Machine Tools and Production Technology (IWF), Technische Universität Braunschweig, Langer Kamp 19b, 38106 Braunschweig, Germany

S. Thiede and C. Herrmann (eds.), *Eco-Factories of the Future*, Sustainable Production, Life Cycle Engineering and Management,
https://doi.org/10.1007/978-3-319-93730-4_9

often operated inefficiently. The holistic approach described in this article suggests specific improvement strategies and innovative technology solutions based on profound evaluations in industrial cases.

9.1 Introduction and Scope

A look at the power demand of industry in Europe emphasizes the role of compressed air. In 2009, the industry sector accounted for a total energy demand of around 970 billion kWh (Eurostat 2009). Especially, the area of actuators and drive technology is relevant, and more than 60% of the industrial electricity demand is attributable to the provision of mechanical power, including about 63 billion kWh alone for the compressed air supply (IEA 2009). Many companies nowadays see a need for immediate action, especially in regard to rising energy prices and increased environmental awareness. Already, simple measures help to reduce the energy demand and significantly reduce energy costs. However, existing potentials are often unused.

In order to improve an existing compressed air system in terms of energy efficiency, the identification of most relevant energy losses and the selection of suitable improvement strategies are major tasks. In Fig. 9.1, the schematic sketch of a compressed air facility and their roughly estimated saving potentials within four main areas (air generation, air preparation, air distribution and air utilization) is shown. This holistic perspective is necessary since losses such as leakages and pressure drops within the upstream infrastructure mainly influence the efficiency of the overall compressed air system. Therefore, this article summarizes identification and improvement strategies for the overall compressed air infrastructure following a systematic approach.

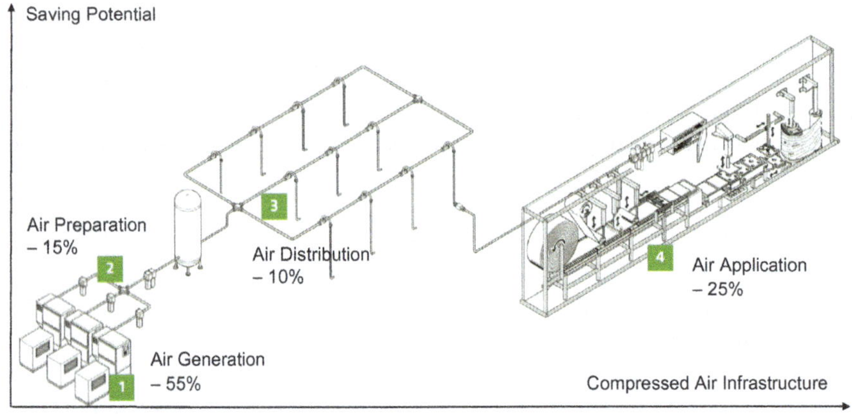

Fig. 9.1 Distribution of saving potentials within the compressed air infrastructure (Hirzel and Köpschall 2012)

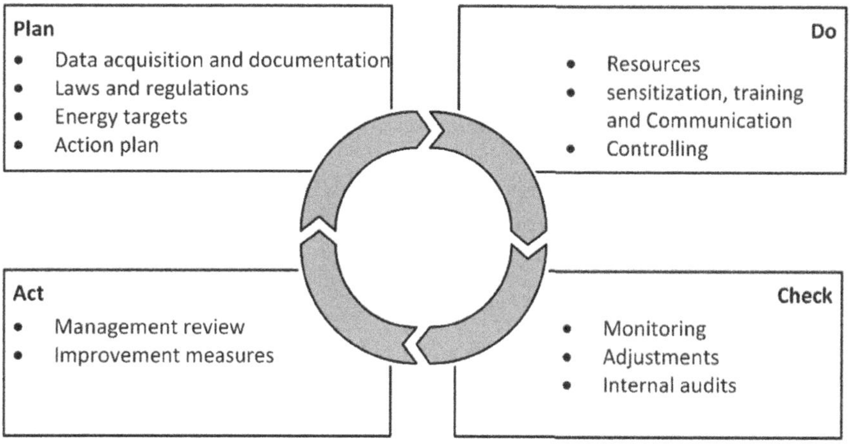

Fig. 9.2 PDCA—process with the four basic tasks and their key activities Adapted from (BMU 2012)

9.2 State of the Art in Energy Auditing in the Field of Compressed Air

Within facility management and the operation of factories, energy management gained importance during the last decades. In the meantime, systematic and normed processes like described in the ISO 50001 are more and more common in the industry. It is used for the systematic increase in energy efficiency in industrial production environments. The EN 15900 on energy efficiency services (DIN 2010) is another systematic approach, which aims at increasing the energy efficiency within the industry. The basis of both approaches is depicted in Fig. 9.2 as an improvement process which can be divided into four basic tasks.

The process starts with the identification of typical energy wastes, followed by the derivation of improvement strategies and the implementation of appropriate measures. Finally, the verification and sustaining of suitable measures guarantee long-term saving effects.

As the given approaches within ISO 50001 and EN 15900 focus on a general management level respectively consider energy efficiency as a whole, important aspects for the applicability within compressed air systems are not described in detail.

9.3 Approach

The nucleus of the developed approach follows a continuous improvement process and can therefore be aligned with processes like proposed in ISO 50001 and EN 15900. Combined the approach includes four steps. The first step targets the identification of energy wastes in compressed air systems. Referring to the identification, improvement strategies are shown. Subsequent the implementation of measures and their verification are described.

9.3.1 Identification of Energy Wastes

Table 9.1 highlights typical losses. The identification of sources for energy losses is explained in the following subsections, where additional strategies for avoiding the detected losses are presented.

Each loss is related to either surplus flow rate or surplus pressure level and can be directly converted into the energy demand at the compressor station. Leakages as flow rate losses have to be compensated by more compressed air generation whereas pressure losses have to be compensated by increasing the pressure level at the compressor station. As a rule of thumb, it can be stated that the increase in 1 bar at the compressor station leads to an increase in energy demand about 6%.

9.3.2 Compressed Air Generation

Compressors are the first element in compressed air systems. Losses in terms of surplus electricity demand are often resulting from wrong dimensioning of compressor stations and from wrong or inefficient supervisory control algorithms.

A suitable strategy for identifying losses within existing compressor stations is the measurement of the demand profile. The measurement has to happen over a sufficient long period of time, i.e. several days for capturing downtime and weekends with basic and peak loads as well as fluctuations in demand. Parameters which need to be

Table 9.1 Main loss sources within the compressed air infrastructure

Air generation	Air preparation	Air distribution	Air utilization
• Supervisory control of compressors • Dimensioning of compressors • Pressure level in compressor station	• Level of air quality • Pressure drops in filters, dryers • Choice of components, e.g. drying equipment	• Pressure drops in pipes • Leakages	• Dimensioning of components • Leakages • Use of sealing air, blow air

captured for further loss detection are as follows: compressor operating times, energy demand measurements (current and voltage of each compressor), compressed air flow measurement (with analogous flow sensor) and pressure levels within compressor stations as well as required range of pressure for control at compressors.

The conduction of measurements helps to improve:

- Costs transparency of compressed air demand: With the measurements of energy demand of compressors and preparation equipment as well as the measurement of supplied compressed air volume with analogous sensors, the costs of compressed air demand can be calculated with *energy intensity of produced compressed air* [kWh/Nm^3] = *total electrical energy demand* [kWh]/*supplied air volume* [Nm^3] and the prices for electrical energy. As average values 2.0–2.5 cent/Nm^3 are given in practice, but values have in general to be checked and determined for each individual industrial set-up. As further reference, the energy intensity should be at about 0.1 kWh/Nm^3 at a given pressure level of 8 bar (Gloor 2000).
- Transparency of energy demand of the entire system: The measurement of compressor operating times and the pressure bandwidth at the centralized storage offer possibilities to put transparency into the process of compressing air. Several things can be detected such as performance of control of compressors (ratio between idle time and run-time of each compressor and overall station, degree of capacity utilization of compressors, matching of compressor power to required flow output). Furthermore, information about output reserves of the system is available.

9.3.3 Compressed Air Preparation

Compressed air preparation must be capable of supplying compressed air at precisely defined quality. Since oil, water or particles in compressed air have a negative effect on the service life span of pneumatic components, the air quality needs to be considered. They cause washout of lifetime lubrication and increased wear and damage to gaskets. Energy and operating costs rise; in worst case, it can lead to unexpected production downtime. Losses are often resulting from excessive requirements regarding the air quality or the use of outworn equipment.

A suggested strategy for identifying losses within existing air preparation facilities is measurement for compressed air quality. Parameters like the pressure dew point and oil concentration are needed to be captured for further loss detection and should be measured centralized and decentralized.

This strategy aims at keeping the quality of compressed air at an optimum level: the frequent inspection of centralized air preparation equipment helps keeping the quality of compressed air at an optimum level. Maintenance engineers often install surplus filters and dryers just to keep the air as clean as possible. If high quality of air is not required for the application, surplus equipment needs to be removed to reduce pressure drops (filters) and the usage of cleaning air (dryers).

9.3.4 Compressed Air Distribution

Optimal leakage management radically lowers the cost of compressed air, as leaky components are a great waste of energy and money. According to Radgen and Blaustein, about 16% of the total potential savings can be achieved in this area with the largest potential efficiency gain by just a single type of measure (Radgen and Blaustein 2001, p. 52). In contrast, in the EnEffAH study, Hirzel and Köpschall estimate the saving from regular leakage reduction measures with a 5% contribution to the total efficiency gain (Hirzel and Köpschall 2012, p. 34). In fact, leakages do not occur within widespread compressed air piping systems but within the tubing of the compressed air applications. The losses that are to be identified within the compressed air distribution are air flow losses due to outworn tubing or fittings as well as not properly closed tube endings and pressure drops[1] due to wrong dimensioning of pipes and tubes.

The following strategies for identifying losses within the distribution network are suggested:

(1) Measurement of leaks in the entire plant by either using hand-held ultrasound detectors for manual leakage detection or installing an intelligent leakage detection system.
(2) Centralized and decentralized measurement of pressure, e.g. at centralized storage as well as spot tests at supply units of applications with analogous pressure sensors.

The conduction of measurements helps to improve:

- Categorization and recording of leakages: With the help of ultrasound detection devices, leakages can be detected during operation that means no production downtime is required. Leaks can be rapidly identified and clearly categorized through marking. The classification of leaks helps the calculation of air losses in [Nm^3] or [€] for further cost-benefit analysis.
- Check for complete compressed air system, from compressor to pneumatic application: Test spot measurements of pressure within the overall compressed air infrastructure help the recording of information needed for repairs and improvements. As a rule of thumb, the pressure drop within the main piping system should be less than 0.01 bar (between storage and main piping system) and 0.03 bar (between main piping system and connecting pipe to application).

[1]Pressure drops will only influence the energy efficiency of a compressed air system negatively if they occur between the centralized storage and the valve at the pneumatic cylinder that means within the piping part that is constantly ventilated with compressed air. The pressure drops have to be compensated by an increased pressure level at the generation unit in order to guarantee a minimum supply pressure at the application. This results in higher costs for air generation. The pressure drops within tubing between valve and actuator are similar to resistances and influence the dynamic behaviour significantly.

9.3.5 Compressed Air Utilization

Knowing the air demand of each machine is a vital prerequisite for optimal sizing of the compressed air supply and distribution as well as for determining leakage losses. But the application's demand profile is not only required for the dimensioning of the generation part. It can and should be used additionally to optimize the performance of the application itself—according to the saying "if it's not used, it doesn't have to be generated".

The following strategies for identifying losses within existing applications are suggested:

1. Measurement of compressed air flow profile with the help of analogous flow sensors in supply unit as well as pressure measurements. The air demand profile should be captured over several hours depending on the production sequence in order to measure all the effects.
2. Measurement of air quality in decentralized supply unit based on the measurement of the parameters such as pressure dew point as well as visual inspection of air preparation units.
3. Measurement of leaks in individual machine with the help of manual ultrasound detectors.

The conduction of measurements helps to improve:

- Identification of inappropriate compressed air usage: The information about potential savings within the compressed air application has to be gained from analysing various characteristics such as: demand per machine cycle and per minute, average and max./min. pressure and max./min. rate of air flow. Therefore, the exact demand during both downtime and operation has to be captured. With the correct information on actual air demand, unnecessary energy demand due to oversupply as well as undesired pressure drops due to undersupply can be identified. The detection of leaks with ultrasound is a possibility to detect small leakages during operation.
- Optimal configuration of compressed air supply to the machine: Continuous maintenance of compressed air preparation favours the efficiency of the overall system, such as an increased service life of pneumatic components due to adjusted air quality. Similar to the optimization of centralized filters and dryers, the proper equipment has to be chosen for the decentralized air preparation in order to avoid surplus pressure drops and reduce demand.

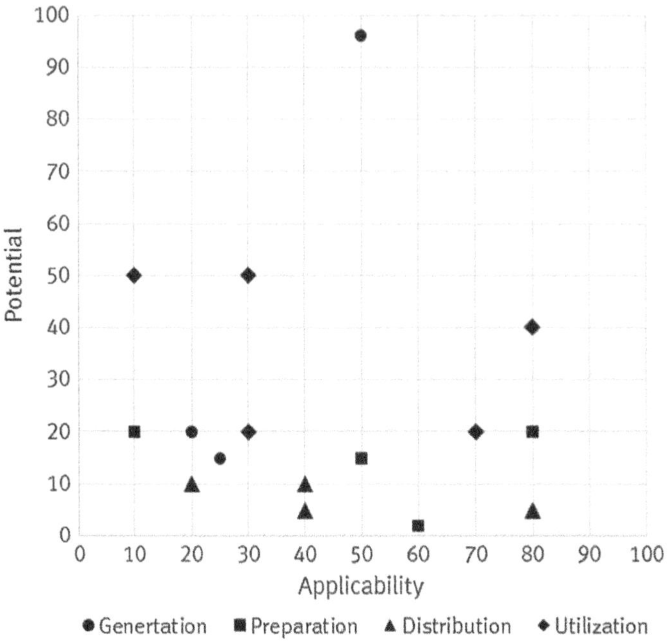

Fig. 9.3 Overview over improvement strategies with typical applicability and saving potentials

9.4 Improvement Strategies

In the following, suggested improvement strategies[2] are summarized for the four main areas of air generation, preparation, distribution and utilization. Figure 9.3 shows a portfolio of all improvement strategies with typical values for their applicability and saving potentials.

9.4.1 Compressed Air Generation

Current, voltage, pressure and flow rate measurements in centralized compressor stations are used to locate loss sources during the process of generating compressed air. Four main improvement strategies are suggested in the following.

- **Dimensioning of compressors adjusted to the demand profile (app. 20 %, pot. 20 %)**: Optimized compressor stations are controlled by an overall control unit which accounts for changes within the demand profile. Several smaller compres-

[2]The improvement strategies are result of the research project EnEffAH (Hirzel and Köpschall 2012). The abbreviation "app." defines the degree of applicability of the shown improvement method; "pot." shows the possible saving potentials. Both values are rough approximations.

sors as well as a mix of frequency-controlled and fixed compressors add to the flexibility of modern stations.

- **Centralized heat recovery (app. 50%, pot. 96%)**: The usage of centralized heat recovery is one of the most promising improvement strategies for compressed air installation. The amount of energy used for generating compressed air is not lowered, but the emitted heat is recovered and used as process heat (water, air) within the company and to support traditional heating.
- **Use of frequency-controlled compressors (app. 25%, pot. 15%)**: The use of frequency-controlled compressors should be always combined with fixed compressor (no speed control). They can be perfectly used for balancing peaks within demand profiles. It should be mentioned that the efficiency of frequency-controlled compressors drops significantly beyond their operation point.
- **Reduction of main pressure level (app. 50%, pot. 15%)**: The reduction of main pressure level offers high saving potentials since it directly relates to the energy demand at the compressors. A reduction is often possible as soon as surplus pressure drops have been avoided.

9.4.2 Compressed Air Preparation

Within the area of compressed air preparation, high pressure drops in outworn components or the too restrictive air quality requirements are offering high saving potentials. Three main improvement strategies are suggested in the following.

- **Choice of drying technology (app. 50%, pot. 15%)**: The choice of technology for drying the compressed air up to a defined dew point influences possible saving potentials. Since refrigerant-type dryer (minimum dew point 3 °C) requires only 2% of the overall energy for generating compressed air, adsorption dryers (minimum dew point −80 °C) use 30%.
- **Dimensioning of air preparation unit (refrigerant-type dryer: app. 60%, pot. 2% and adsorption dryer: app. 10%, pot. 20%)**: Often, too restrictive requirements on the air quality add up to an over-dimensioning of air preparation equipment. This fact results in surplus pressure drops—mains pressure level has to be increased to compensate these pressure drops.
- **Regular check of air preparation equipment (app. 80%, pot. 20%)**: The regular check of air preparation equipment is a relevant factor for guaranteeing the efficiency of the overall system. There can be high and surplus pressure drops in filters that are blocked by oil and dirt. A half-yearly check of filters can end up in a reduction of the main pressure level by 1 bar. This results in a reduced energy demand of approximately 6% due to the reduced pressure level at the compressors.

9.4.3 Compressed Air Distribution

The correct dimensioning of the piping system influences the overall efficiency and is mainly dependent on two loss sources—pressure drops due to wrong dimensioning and leakages due to missing maintenance. Hence, four improvement strategies are suggested:

- **Optimized dimensioning of piping infrastructure (app. 20 %, pot. 10 %)**: Piping systems are often undersized. The reduced pressure level at the compressed air utilization has to be compensated by increasing the main pressure level at the generation unit. The pipes should better be oversized than undersized.
- **Optimization of network topology (app. 40 %, pot. 5 %)**: The choice of network topology decreases or increases the performance of the piping infrastructure. Whereas branched networks suffer from constantly increasing pressure drops, networks with closed circuits pressure drops can be balanced and the absolute pressure level stabilized.
- **Placement of decentralized storages (app. 40 %, pot. 10 %)**: Decentralized storages are a cost-efficient way of stabilizing pressure oscillations in piping networks. They are frequently used close to consumers with high dynamics and huge demand.
- **Reduction of leakages (app. 80 %, pot. 5 %)**: Leakages have to be reduced (refer Sect. 9.5).

9.4.4 Compressed Air Application

Typical machinery for automation processes consists of several components that are using compressed air in different ways. For a better description of relevant improvement strategies, compressed air utilization is divided into five categories such as vacuum applications (1), use of blow and sealing air (2), grippers (3), pneumatic cylinders (4) and pneumatic pivot drives (5). The following improvement strategies for the aforementioned five application types are highlighted.

- **Reduction of pressure level applicable to types (1) to (5)**: The pressure at the decentralized supply unit (right before each application) should be reduced up to the point where the machinery does not function anymore. This avoids over-dimensioning. In case of modern machines with correct dimensioning, this improvement strategy is not applicable!
- **Replace double-acting cylinders by single-acting ones for types (3) and (4) (app. 10 %, pot. 50 %)**: A spring is used for realizing the return stroke of the cylinder. Hence, half the amount of air is saved within one stroke.
- **Reduce size of components applicable to type (1) to (5) (app. 80 %, pot. 40 %)**: The dimensioning of automation systems that means the choice of components in type and size should always be done by experts. There are tools for optimized dimensioning available by component suppliers. The correct dimensioning is the most important step towards an efficient drive system.

- **Reduce pressure level for unproductive return strokes applicable to type (3), (4) and (5)**: In case of unproductive return strokes, the pressure level can be reduced.
- **Shorten tubing length applicable to types (1) to (5)**: As stated earlier, the size of piping between valve and actuator influences the air demand of the drive system. They represent a dead volume that needs to be filled within each stroke. Hence, tubes should be shortened as much as possible. Additionally, the diameter accounts for the air demand. Tubes with too small diameters function as resistance and influence the dynamics negatively.
- **Air-saving circuits (app. 30%, pot. 50%)**: There are several possibilities for the implementation of air-saving circuits, e.g. for usage in vacuum generators. Nevertheless, the shortening of running time of blow and sealing air components is a simple but effective improvement method.
- **Avoidance of dead volume (app. 30%, pot. 20%)**: Especially in cylinders and tubing, dead volume should be reduced as much as possible.
- **Elimination of leaks in application (app. 70%, pot. 20%)**: Most leakages occur inside machinery and only a few within the overall piping infrastructure. Above all, outworn and leaky fitting cause flow losses.

9.5 Implementing Measures

The implementation of appropriate measures calls for elaborated concepts. In the following, the most relevant measures to improve energy efficiency along the whole compressed air infrastructure are described.

9.5.1 Optimization of Parameters in Compressed Air Generation and Preparation

The measurements and correct interpretation of the parameters should be in general done by specialists for compressed air installations. These service providers often offer trainings for customers such that they can do these measurements on their own and/or derive the necessary plans of action.

9.5.2 Leakage Management System

The detection of leakages with ultrasound equipment is comparably time-consuming in typical industrial plants. A leakage management system is according to the Plan-Do-Check-Act (PDCA) management cycle supporting the continuous and automatic monitoring of compressed air systems. There are three different levels of leakage management suggested (dependent on flow volume and power of compressed air installation in factory)—Level 0: One-time manual leakage detection and repair typically done by external service provider, e.g. for smaller outworn compressed air installations; Level 1: Yearly manual leakage detection by trained employees as well as yearly check by maintenance service for compressors; and Level 2: Installation of distributed sensor equipment in piping system (e.g. at storage, junctions in main piping system, at supply unit of each application) and automatic monitoring by overall Supervisory Control and Data Acquisition (SCADA) system with alert function (e.g. changes in demand profile, increase in pressure drops). The leakage management concepts in Level 1 and Level 2 have to be defined as a personal action plan for savings, with systematic documentation of leak elimination. Action plans have to be defined for rapid and lasting leak elimination.

9.5.3 Condition Monitoring and Diagnostic Systems for Application

Condition monitoring and diagnostic systems help to detect changes in pressure and flow rates early and to prevent impending production downtime. If downtime still occurs, they enable rapid detection of the problem. A condition monitoring system has to fulfil the following objectives such as: continuous monitoring of critical machine processes, prevention of unplanned downtime as well as permanent demand monitoring and rapid detection of deviations. It supports the maintenance engineer during the analysis of the process and the identification of the parameters to be monitored. It furthermore saves the parameters. An elaborated condition monitoring system guarantees energy-optimized system operation, transparency of demand, constant production quality, maximum availability and budgeting for repairs.

9.5.4 Improvement of Existing Machinery

Improvement concepts for compressed air applications should obviously go beyond fixing leakages. Further savings by strategically redesigning the compressed air system for the future have to be addressed. This will not only stabilize the processes but also reduce the costs. Based on the previous losses detection and improvement strategies optimization, measures for relevant machines need to be implemented. For the analysis of relevant applications in a machine, two main phases are suggested, namely a general and detailed analysis which are shown in Fig. 9.4. Before starting the analysis, a comprehensive picture about relevant data such as the energy demand and expected cost-effectiveness should be available (initial phase). The main first phase as the general analysis allows a structured analysis of the components to

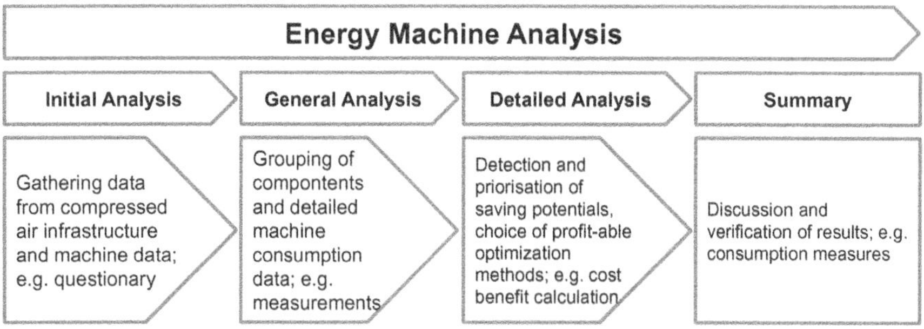

Fig. 9.4 Main phases of the machine optimization concept

identify compressed air savings. The detailed analysis as the second phase determines the actual demand and the impact of the specified measures to optimize the demand of the components. This can be the basis for the economic evaluation of optimization measures and further recommendations. Finally, the optimization has to be performed either by service experts or by trained employees. Within the summary phase, the verification of implemented optimization methods has to be done by evaluating measurements.

The optimization of existing machinery aims at the reduction of energy and compressed air demand as well as the shortening of cycle times, simplified installation, simplified maintenance and improved process reliability.

9.5.5 *Engineering of New Machinery*

Instead of optimizing an existing machine, the correct dimensioning and engineering of a new machine offers far more saving potentials also in case of drive systems. The choice of pneumatic cylinders with or without rod as well as the choice of cylinder diameter and damping technology influences the dynamics of the system as well as the energy demand. The combination of "technology choice", "operational strategy" and "dimensioning" leads to the implementation of a modern and energy-efficient actuation system. Most manufacturers for drive components offer engineering tools supporting the customers during the process of machine engineering. Axle-type drive components represent a small number of components since they can vary in diameter, stroke and guide. The process of component choice is challenging and requires expert knowledge.

The main steps during this process are as follows: the initial step is the definition of the application and specification of requirements such as moving/gripping and installation position. Relevant limitations such as existing infrastructure required safety classes and environmental conditions are to be defined as well. With the help of simulation software, a pre-assessment of components can be made respecting defined technical criteria such as geometry aspects, accuracy limitations, dynamics and mechanical limitations such as guides. A set of possible solutions is the results of the pre-assessment and needs further design criteria for the final choice of components. Final selection criteria are stock availability of components at the customers, effort for maintenance, life cycle costing (LCC) calculations as well as existing technology know-how at customer.

9.6 Verification and Sustaining

9.6.1 Verification

Following the implementation, the improvement measures have to be verified. After the improvement measures have been implemented, a further measurement can determine the success of the actions taken. This can be done easily by performing the same measurements as defined for the identification process. Finally, a sustaining process has to be established in order to guarantee that losses are avoided in advance and errors during design and operation of machinery are not to be repeated.

9.6.2 Sustaining

Section 9.5.2 described three different levels of leakage management. A professional elimination of leaks, i.e. the repair or replacement of leaking or faulty pneumatic components, has to be performed in order to stop wasting energy. This often provides the earliest return on investment. Looking at long-term benefits, the mentioned activities must be taken regularly to maintain the optimized condition of the machines and the compressed air infrastructure.

The benefits from using external expert services are proven. With the help of external maintenance services (ideally compressed air or pneumatics specialists), the systems can be brought to the latest energy-saving standards. Typical preventive and corrective maintenance services include:

- Inspection: Checking for damage, component inspection, checking of air preparation and filtration.
- Preventive maintenance: Air filter replacement, silencer replacement, lubrication/relubrication of guides, etc., tightening of loose fittings, screws and belts.
- Corrective maintenance: Elimination of leaks, component replacement, component repair.

In a second step, employees' awareness of energy efficiency should be raised and barriers must be reduced. This is essential to ensure that the measures implemented are successful in the long term. This newly raised awareness should also be communicated to suppliers (e.g. in performance specifications) and customers (e.g. via marketing).

The approach was evaluated at partner facilities of the automotive and rail industry sector focusing on their specific needs (Chap. 4). The involved partners covered a broad range of competences and interests such as technology provider, system integrator, service provider and manufacturer. These different perspectives and contributions are relevant to identify and implement the most promising optimization potentials.

9.7 Application

The analysis at the partner's facilities increased the level of transparency and clearly indicated the hot spots in terms of areas with high compressed air demand levels. Audits with mobile inspection and measurement equipment such as ultrasound detection devices provided an overview of occurring leakages in relevant application areas, e.g. on machine level between valve and the clamping devices.

In the following chapters, chosen solutions for specific applications are described.

9.7.1 Supporting Technology Decision in Engineering of Actuation Systems in Automotive Industries

Same as in every other industry, production processes in the automotive industry are based on a number of various actuation tasks. Figure 9.5 shows a selection such as gripping of car assembly parts, positioning of cutters and applying high forces for pressing processes. Which technology to choose is first of all a question of fulfilling technical specifications such as cycle time, forces and loads. Especially in automation, one actuation task might be realizable with different technologies (e.g. electrical, pneumatic and hydraulic) and nowadays with many components in different sizes and power ratings. That degree of freedom offers the chance for an energy-optimized design of actuation systems during the machine planning process, namely the use of an assessment tool.

Within the shown case, a simple but every effective set-up for an assessment tool for energy efficiency improvements within the application scenarios is identified, while quantifying the potential energy savings as well as payback time for investments and maintenance costs at the same time. This way, decision-makers have the possibility to carry out energy diagnosis focusing on most relevant alternatives and getting reliable data during the planning process at comparatively low costs.

The principle set-up for the assessment is highlighted in Fig. 9.6 and depicted as follows:

- Separation of process in single actuation tasks S

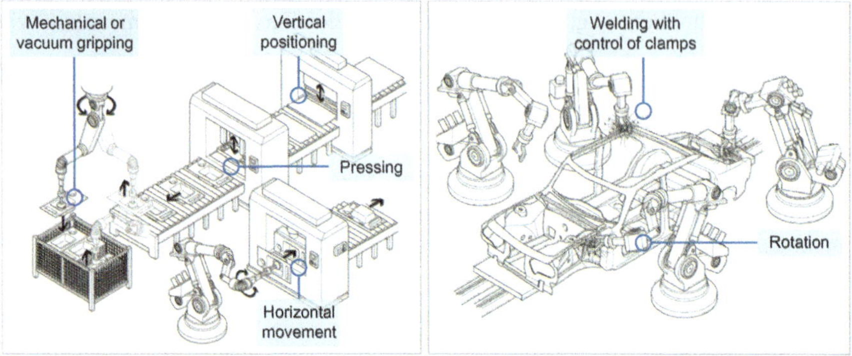

Fig. 9.5 Manufacturing processes in the automotive industry require various automation tasks such as horizontal or vertical movements and positioning, pressing, gripping or high-speed rotations

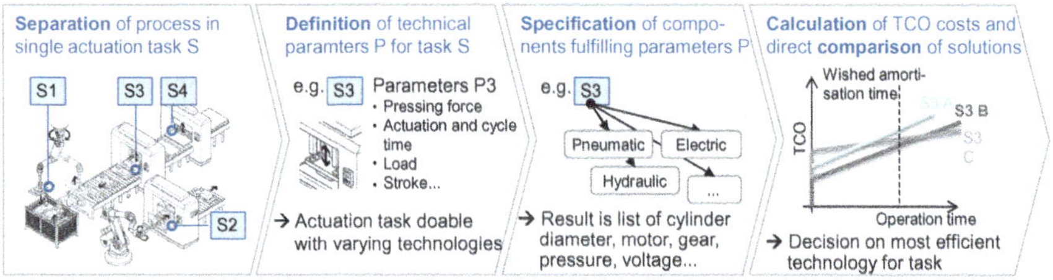

Fig. 9.6 Step-by-step process for supported technology decision for actuation systems in automotive industry

The separation should be done on component level. Several tasks could be later aggregated on a cell or machine level to represent different application scenarios.

- Definition of technical parameters P for each task S
  With a pre-done mapping of major tasks in automotive industry (e.g. gripping, welding, positioning), the main parameters are easily added to each task. Typical values are the load, the stroke length, cycle time and standby time, relevant forces.
- Specification of components fulfilling parameters P
  To be evaluated, components are either given or can be gathered by an additional design tool with dynamic simulation support. The result is a list of component sizes and power ratings fulfilling the technical parameters P as well as investment costs for each configuration. Sometimes, the dynamics simulation support already calculates the energy demand for one cycle which can be used for the next step.
- Calculation of total costs of ownership (TCO).

For complex functions that cannot be described by formulas or empirically, it can be helpful to use simulations, e.g. to determine the energy requirements of electric drive systems in a control loop with different positioning profiles. The graphical and numerical analysis can be performed with a simple break-even point determination. At the end of the assessment, a mapping of the whole hierarchical system structure (e.g. whole production line) with an aggregation of the results on the individual actuation tasks is possible.

On the basis of the derived data, a decision-maker can decide what type of, e.g. clamping device is favourable in the long-term economic sense or at a specific amortization time for the machinery. Analogous, calculations can be done for other components such as welding guns. The more detailed the back-end calculations for energy costs are, the more precise the TCO forecast will be.

The shown way of technology assessment with demand data and break-even point (compare Fig. 9.6) should be used nowadays when discussing of energy-optimized production processes. They give a clear indication about the share of energy costs in each production line and support approaches for energy-transparent machinery. The decision-maker is now able to decide to what extent this information is relevant and may support to switch to other actuation technologies.

9.7.2 Creating Energy Transparency on Machine Level

Energy data is important for a strategic energy management—especially, the identification of potentials as well as the verification of implemented measures depends on the availability of this data. As Müller et al. state a typical challenge in production is this availability of energy data from a top factory level down to the machine level (Müller et al. 2009). Furthermore, bare energy data without additional data or reference values makes a profound evaluation hard or even impossible. Additional data can be machine data (e.g. machine states) or production data (e.g. information about the processed order, the number of produced parts or the currently produced part).

A possible solution to this problem is a continuous energy monitoring system with distributed sensors on machine level, which is able to complete monitoring solutions on building level (Posselt 2015). The following approach focuses on a solution that monitors electrical energy and compressed air demand as well as the machine states of the corresponding machine.

For a first practical implementation, a machine group with an estimated high energy-saving potential was chosen and equipped with suitable sensors for electrical energy and compressed air flows. The machines are part of a job-shop production as it can be found for the production of automation components (e.g. valves and valve terminals). To ensure the retrofit ability, clamp-on current transformers for electrical energy measurement were selected. The compressed air demand is measured using a pressure and a flow sensor.

An additional PLC captures the sensor values and obtains machine states via a data connection to the main PLC of the machine. Both PLCs are placed within the electrical cabinet of the machine. The additional PLC is able to pre-process the energy and machine data and provides a basic visualization. The number of produced parts is not provided by the main control of the machine group. To calculate KPIs like "required energy per part", the produced parts are estimated with a counter for the

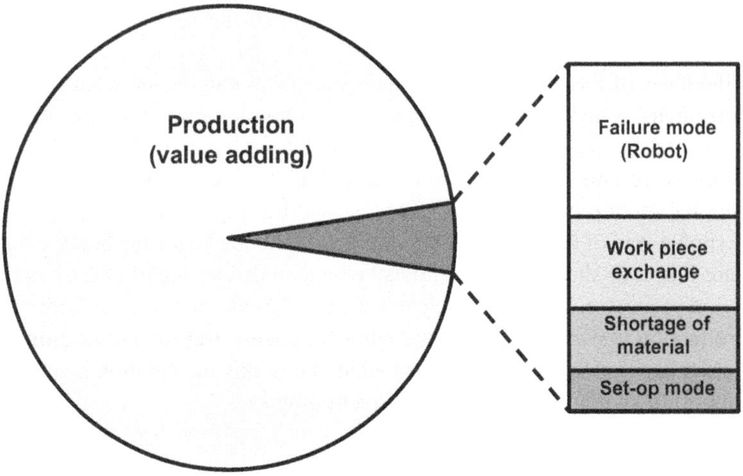

Fig. 9.7 Production state and non-value-adding states of a machine

machine state "changing workpiece". The pre-processed data is made available to supervising systems via OPC UA. Supervising systems can be SCADA, MES, ERP or energy management software systems. Those systems are capable of automatically evaluating the data from the machine. Further functionality could be intelligent alerting, the generation of reports or the visualization of the data for different users.

With the additional equipment, a continuous monitoring of the energy flows on the machine level could be achieved. Energy flows can be separated into value-adding energy, if the machine is in "production state", and non-value-adding energy, in case the machine is in states that are not productive like "standby" or "parts shortages". The base for new benchmarks and KPIs like "used energy per produced (good) part", "energy per processed order" or "percentage of power consumed in value-adding machine states" (see Fig. 9.7) is now given. These benchmarks and KPIs can be used for many purposes like:

- Supporting maintenance work (to identify issues or for predictive maintenance).
- Identifying optimization potentials.
- Tracing actions for increased energy efficiency with respect to their effectiveness, adherence or the achieved savings.
- Allocation of energy costs to different organizational units.
- Raising awareness for energy usage.

In further steps, the developed solution will be transferred to more machines that exceed a threshold in energy demand which makes them relevant for a continuous monitoring. For machines with an energy demand below the threshold or old machines with a low estimated remaining operating time, temporary measurement can support the overall energy transparency. Equipping existing machines with hardware to obtain energy transparency is just a first step towards an energy-transparent factory. The energy-transparent machines together with energy metres from the facility build the basis for an energy transparency system. The energy transparency system has the following main features:

- Providing visualization of all relevant energy flows of a factory.
  - Seamless integration of energy data from facility and production.
  - Adapted visualizations for different user roles and their preformed tasks.
- Support of continuous improvement processes.
  - Identification of improvement possibilities.
  - Verification of improvement measures.
- Supervision of energy flows.
  - Detecting leakages or pressure drops in compressed air networks.
  - Detecting load peaks.
  - Detecting unusual behaviour of machines as a precursor of machine fails.

9.7.3 Intelligent Machine Monitoring in Rail Industry

9.7.3.1 Problem Statement

The guidelines above provide various approaches to optimize compressed air installations and applications in new and existing factory environments, whereas assessing saving potentials of compressed air distribution and utilization systems is a perpetual issue in production environments. Especially, industries with a high degree of manual manufacturing operations profit from stepwise improvement in production, following the well-known lean principles. As an example, the manufacturing of metro cars requires relatively a high share of manually performed operations, focussing on the area of roof and undercarriage assembly. The dominating manual production operations are weld seam preparation (bevelling) and weld seam finishing (trimming). The manual chip-cutting operations are performed with compressed air-driven power tools (angle grinders, vertical grinders, die grinders, drills, tappers and riveting hammers), as exemplarily depicted in Fig. 9.8.

Compressed air-driven tools are favoured over electric tools in the given production site because of the presence of very rough working environments (high dust and noise emissions), comparably lighter tools and very robust mechanics of such. The disadvantage of the utilization of compressed air-driven tools in rough working environments is the fact of seemingly rising leakages due to ageing and contaminated joint and gaskets. Gaskets, couplings and pneumatic hoses are subject to frequent connection and disconnection of power tools, which is accelerating and conveying the contamination and ageing processes. These effects are even hard to reduce by frequent maintenance intervals. Therefore, the degree of leakages in the present working environment is fairly high in comparison with higher automated production facilities.

Fig. 9.8 Pneumatic power tool (grinder) used in manual post-processing of welding seams

Due to the high noise emissions in the working environments, obvious leakages of broken gaskets or joints are hardly perceptible. Hence, leakages are difficult to be removed and are therefore omnipresent.

The operation of an optimized compressed air infrastructure can be further supported by decentralized, automated devices, which shut down the energy supply (compressed air) of supply net branches during non-demand times. This approach is described in the case at hand. To verify and assess the potential of an automated compressed air supply management in production, a functional prototype was tested at Siemens production site in Vienna. The goal was to counter the problem of wear-out effects in train manufacturing. Therefore, a compressed air-saving module was developed, which was placed in the branch lines of the compressed air distribution infrastructure. The branches are leading to manual workstations. After laboratory testing of the module, the smart distribution network was implemented in the pilot demonstrator. The prototype of the air supply unit was placed in the body in white production for train coaches. Manual manufacturing steps, such as milling and deburring after joining processes, are performed by various hand-held compressed air power tools.

The module, depicted in Fig. 9.9, contains a smart control unit which is capable of automatically recognizing demand and standby times of connected power tools. During demand times, the smart distribution network supplies the power tools with compressed air at grid pressure level. During standby times, the grid pressure is auto-

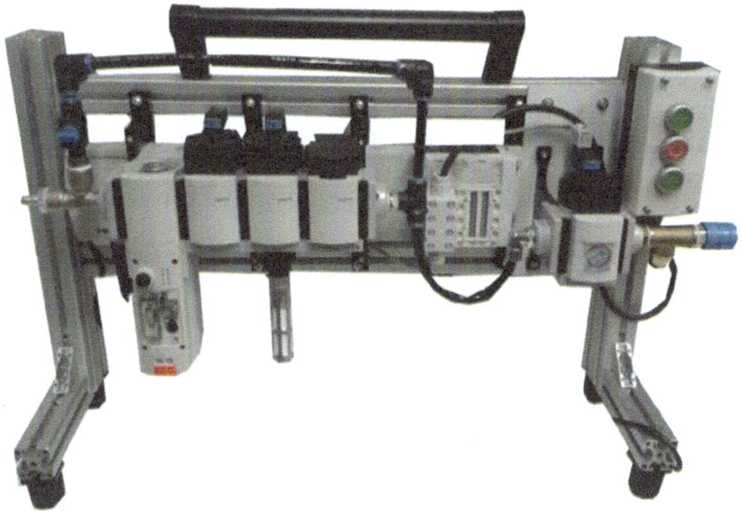

Fig. 9.9 Mobile functional prototype of the smart air-saving unit

matically cut off and leakages in the downstream periphery are therewith reduced to a minimum. When demand occurs again after an automated pressure cut-off, the unit is capable of automatically recognizing the demand and seamlessly resupplying the power tools with full grid pressure. This behaviour can be seen in Fig. 9.10. Additionally, the smart unit is capable of logging productive compressed air demands and the saved amount of pneumatic energy on a continuous basis. For the demonstration case, an additional visualization unit allows to present these performance indicators to the observer on a human–machine interface (HMI screen), as depicted in Fig. 9.12. The testing period was set to 4 weeks of operation. The gained data was evaluated according to the potentials, and qualitative interviews were conducted along the working personnel to estimate the practical influence of the automated shutdown functionality.

9.7.3.2 Functional Prototype

To reveal and assess saving potentials, a modified air supply unit with prototypical functions, as described in Table 9.2, was installed at a compressed air supply hose in a manual manufacturing workstation. The hardware of the module includes sensors to monitor the volume flow and the pressure level, as well as valves for safe switching of the compressed air supply.

The installation was placed in a pipe of the compressed air infrastructure near a manual working station, as shown in Fig. 9.11. The volume in the hose between the module and the coupling was used for implementing the described functions

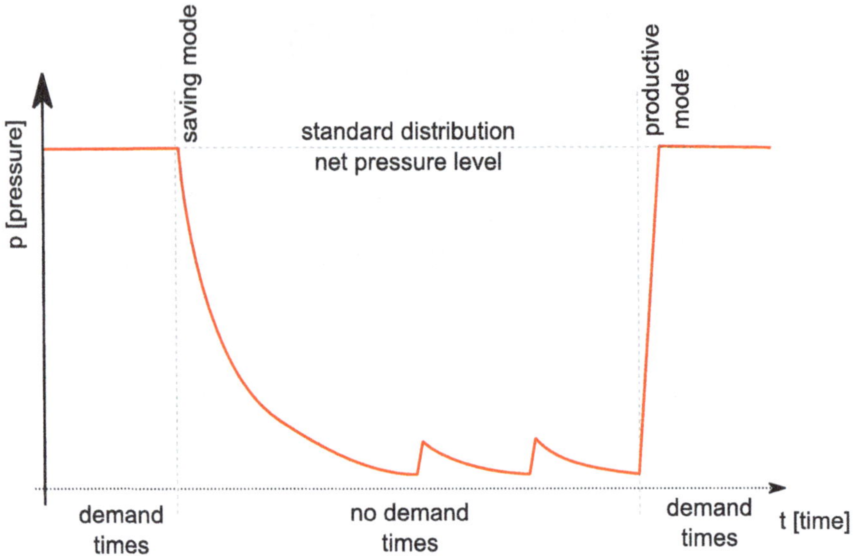

Fig. 9.10 Ideal pressure demand profile of the manual working station

Table 9.2 Additional functions of the compressed air supply unit

Function	Description
Automated shutdown	The module shuts down the energy supply if there is no active use of compressed air
Automated switch-on	The module smoothly switches on the compressed air supply with a short delay if a demanding consumer is detected
Savings monitoring	Based on pressure gradients in the closed supply branch, the leakage rate of the downstream branch is calculated
Estimation of potentials	The potentials are estimated based on the leakages and certain site-specific factors

(compare Table 9.2). For a period of 4 weeks, data was acquired and saving potentials were monitored.

Besides the leakage monitoring, the module is capable of performing network condition monitoring (pressure level, pressure drops, maximum and average demands, etc.). Leakage-level alarms can be triggered visually and as a digital output signal for external alarms processing.

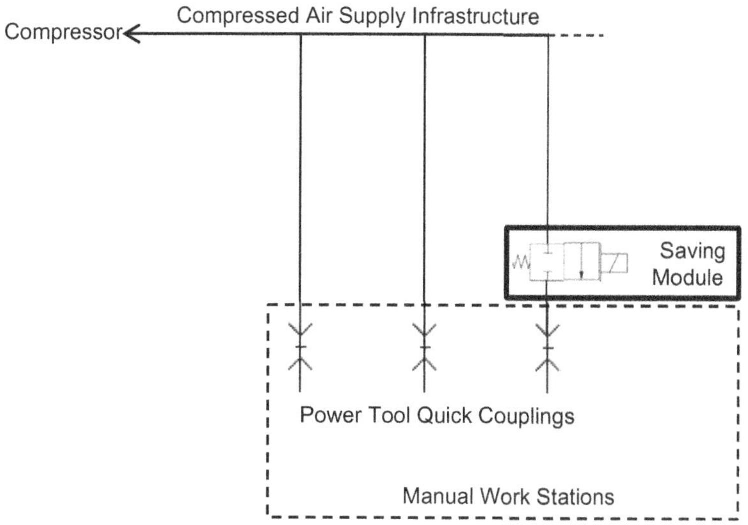

Fig. 9.11 Application environment of the compressed air-saving unit prototype

9.7.3.3 Results

The environmental and economic benefits are potentially high, as leakages during standby times are reduced to a technically possible minimum. During demand times, when power tools are in use, the leakages cannot be reduced automatically, but due to a continuous condition monitoring, alarms can guide energy-saving services by providing an instant return of investment (human workforce and maintenance material) calculation, as depicted in Fig. 9.12, in the screenshot from the detachable HMI screen. The smart distribution network is installed for long-term testing at the demo site to evaluate specific savings and suitable parameter values to improve the control algorithms to be adaptable for varying application conditions.

The assumed high savings were validated by the conducted measurements. The assessment during the testing period leads to high savings and hence quick amortization times below 1 year. A challenge for a broad implementation of such a decentralized solution is the unforeseen malfunctioning of the control unit, as the module is designed to seemingly integrate in the workflows of daily operations. The huge advantage of the developed solution is that shop floor personnel do not need to change their behaviour during operation of the hand-held tools, which favours a high degree of acceptance. Conclusively, retrofitting existing compressed air supply infrastructure to a *Smart Distribution Network* is recommendable for certain industries branches with a high share of utilized manual power tools, as the Siemens case above proves. Furthermore, continuous monitoring of compressed air networks featured by those decentralized devices provides the basis for the initiation of demand-oriented energy-saving services and reactive maintenance approaches in production. Combined with

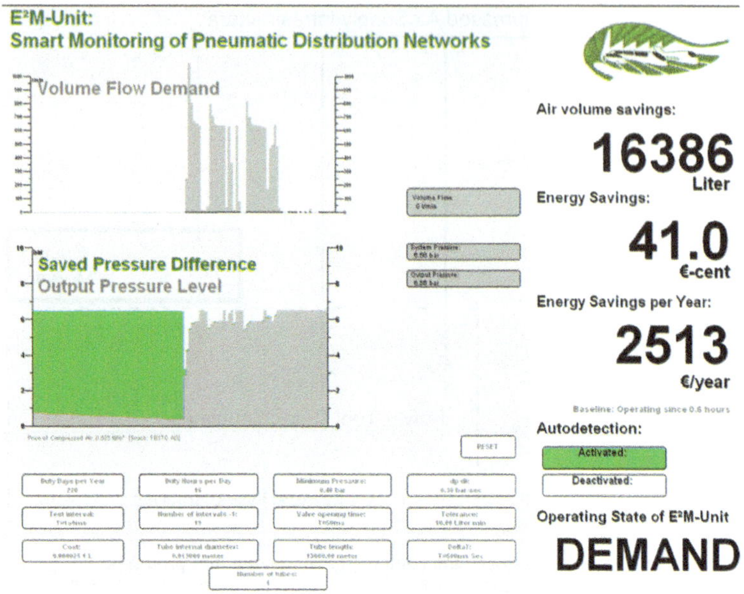

Fig. 9.12 User interface for quick evaluation of the functional prototype in the demonstration environment

advanced control and online analytics, even concepts approaching predictive maintenance become accessible.

9.7.4 Supported Technology Decision for Actuation Systems in Automotive Industry

Same as in every other industry, production processes in the automotive industry are based on a number of various actuation tasks. Figure 9.5 shows a selection, such as gripping of car assembly parts, positioning of cutters and applying high forces for pressing processes. Which technology to choose is first of all a question of fulfilling technical specifications such as cycle time, forces and loads. Besides that, and especially in automation, one actuation task might be realizable with different technologies (e.g. electrical, pneumatic and hydraulic) and nowadays with many components in different sizes and power ratings. That degree of freedom offers the chance for an energy-optimized design of actuation systems during the machine planning process, namely the use of an assessment tool.

Within the EMC2-Factory project, a simple but very effective assessment tool for energy efficiency improvements within was developed, able to quantify the potential energy savings as well as payback time for investments and maintenance costs at the same time. This way, decision-makers have the possibility to carry out, with low cost, a thorough energy diagnosis focusing on most relevant alternatives and getting reliable data during the planning process.

The principle set-up for the assessment is highlighted in Fig. 9.6 and depicted as follows:

- **Separation** of process in single actuation tasks S
  The separation should be done on the component level. Several tasks could be later aggregated on a cell or machine level to represent different application scenarios.
- **Definition** of technical parameters P for each task S
  With a pre-done mapping of major tasks in automotive industry (e.g. gripping, welding, positioning), the main parameters are easily added to each task. Typical values are the load, the stroke length, the cycle and standby time and relevant forces.
- **Specification** of components fulfilling parameters P
  To be evaluated, components are either given or can be gathered by an additional design tool with dynamic simulation support. The result is a list of component sizes and power ratings fulfilling the technical parameters P as well as investment costs for each configuration. Sometimes, the dynamics simulation support already calculates the energy demand for one cycle which can be used for the next step.
- **Calculation** of TCO costs (investment and energy costs)
  A connection to backend simulation for those complex functions which cannot be described through formulas or empirically might be helpful, e.g. energy demand of electric drive systems in a feedback control-loop with different positioning profiles. The graphical and numerical analysis can be performed with a simple break-even point determination. At the end of the assessment, a mapping of the whole hierarchical system structure (e.g. whole production line) with an aggregation of the results on the individual actuation tasks is possible.

On the basis of the derived data, a decision-maker can decide what type of, e.g. clamping device is favourable in the long-term economic sense or at a specific amortization time for the machinery. Analogous calculations can be done for other components such as welding guns. The more detailed the back-end calculations for energy costs are, the more precise the TCO forecast will be.

The shown way of technology assessment with demand data and break-even point figures (compare Fig. 9.6) should be used nowadays when discussing energy-optimized production processes. They give a clear indication about the share of energy costs in each production line and support approaches for energy-transparent machinery. The decision-maker is now able to decide to what extent this information is relevant and may support to switch to other actuation technologies.

9.8 Results and Outlook

The described approaches provide a guideline for operational energy saving in the area of compressed air. It has been developed together with the involved partners and evaluated at partner's production sites as a part of the European research project EMC2-Factory.

The implementation of the developed solutions as described in Sect. 9.4 has shown that this has to be done under consideration of the specific needs, based on an iterative analysis and covering a wide section of the live cycle. It begins with the consideration of energy efficiency during the design and engineering, covers the operation phase with new approaches for the energy monitoring during the operational phase and ends with closing the control loop by automatically switching of parts of the compressed air distribution in unproductive times to make use of identified energy-saving potentials.

The described measures and solutions are rather to be understood as possible blueprints and patterns than precise requirements. Each industry application has its specific boundary conditions and calls for more or less tailored solutions which should be developed together with partners.

The research work in the field of using compressed air more efficiently is ongoing, and more innovative solutions are on the way. It is foreseeable that drive components have to become smarter, e.g. to monitor or control the energy demand more effectively and in correlation to the real demand of the desired application in terms of technical parameters (such as force or velocity).

Acknowledgements The research leading to the presented results has received funding from the European Community's Seventh Framework Program (FP7/2007-2013) under the grant agreement no. 285363 with the title "Eco Manufactured Transportation Means from Clean and Competitive Factory" (Posselt et al. 2013). For further information about the project, visit www.emc2-factor y.eu.

References

Bundesministerium für Umwelt, Naturschutz und Reaktorsicherheit (2012) Energiemanagementsysteme in der Praxis - ISO 50001: Leitfaden für Unternehmen und Organisationen

DIN (2010) Energieeffizienz-Dienstleistungen – Definitionen und Anforderungen; Deutsche Fassung EN 15900:2010

Eurostat (2009). http://ec.europa.eu/eurostat/help/new-eurostat-website

Gloor R (2000) Energieeinsparungen bei Druckluftanlagen in der Schweiz

Hirzel S, Köpschall M (2012) Energieeffizienz in der Produktion im Bereich Antriebs- und Handhabungstechnik, Project booklet. http://www.eneffah.de

IEA (2009) International Energy Agency (IEA). http://www.iea.org/

Müller E, Engelmann J, Löffler T, Strauch J (2009) Energieeffiziente Fabriken planen und betreiben, Springer. http://link.springer.com/book/10.1007/978-3-540-89644-9, p 288

Posselt G (2015) Towards energy transparent factories. Springer International Publishing. ISBN 978-3-329-20868-8

Posselt G, Linzbach J, Weinert N, Kaluza A, Slupek K, Bernas M, Palasciano C, Eytan A (2013) the eco-factory solutions for experts and decision makers project booklet. http://www.emc2-fact ory.eu

Radgen P, Blaustein E (2001) Compressed air systems in the European Union. ISBN 3.932298-16-0

VDMA (2014) VDMA-Einheitsblatt Nr. 66412-10, Manufacturing Execution Systems—Daten für Fertigungskennzahlen. http://www.vdma.org/documents/105628/2244593/Entwurf+VDMA+66 412-10

The manufacturer's authorised representative in the EU is Springer Nature Customer Service Centre GmbH, Europaplatz 3, 69115 Heidelberg, Germany. If you have any concerns regarding our products, please contact ProductSafety@springernature.com

Printed and bound by CPI Group (UK) Ltd, Croydon, CR0 4YY
01/10/2026
02225975-0004